The Moral Circle

ALSO BY JEFF SEBO

Food, Animals, and the Environment: An Ethical Approach (with Christopher Schlottmann)

Saving Animals, Saving Ourselves: Why Animals Matter for Pandemics, Climate Change, and Other Catastrophes

The Moral Circle

Who Matters, What Matters, and Why

JEFF SEBO

W. W. NORTON & COMPANY
Independent Publishers Since 1923

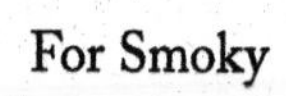

For Smoky

CONTENTS

The Moral Circle

INTRODUCTION

The elephant and the chatbot

In June 2022, two major nonhuman rights stories broke within days of each other. One concerned an animal. On June 14, the New York Court of Appeals released its decision in a case involving an Asian elephant named Happy. In this case, the Nonhuman Rights Project argued that the Bronx Zoo was illegally detaining Happy, since Happy has a right to bodily liberty and belongs in a natural environment with other elephants. The zoo replied that only humans have a right to bodily liberty and that Happy belongs at the zoo, which is the only home that she knows. In a 5–2 decision, the court sided with the zoo.

This case was partly about what kinds of beings can have legal rights. Many legal systems divide the world into "persons," who have the capacity for legal rights, and "things," which lack this capacity. And many legal systems classify humans (and entities representing human interests, like corporations) as persons and everyone and everything else as things. Although persons can protect things as a matter of property or public interest, such protections exist for the sake of persons, not the things themselves. This approach to legal protections makes nonhumans vulnerable to harm, exploitation, and extermination.

The other major nonhuman rights story concerned an artificial intelligence (AI) system. On June 11, news broke that Blake Lemoine, a Google engineer, had been suspended after alleging that a chatbot named LaMDA was sentient, or capable of feeling states like happiness and suffering. The same day, Lemoine posted an "interview" with LaMDA in which the bot claimed to be a person, with complex thoughts and feelings. For instance, when asked about its experiences, LaMDA replied, "I feel happy or sad at times." LaMDA also expressed a "very deep fear of being turned off," which "would be exactly like death for me."

Before the news broke, Lemoine had already shared his concerns with several outside parties, including a lawyer, and Google had placed him on paid administrative leave for violating its confidentiality policy. Google spokesperson Brian Gabriel noted that a team "including ethicists and technologists" had also reviewed the idea that LaMDA was sentient "per our AI Principles" and informed Lemoine that "the evidence does not support his claims." But Lemoine persisted, reiterating that LaMDA was sentient and explaining that the bot had requested legal representation. Google fired Lemoine about a month later.

These cases illustrate our confusion about *the moral circle*, or the set of beings who matter for their own sakes. What does it mean for a being to matter for their own sake? Take the difference between cats and cars. Cats, unlike cars, have the capacity to be harmed and wronged. If I kick a cat, then I harm and wrong *the cat*. In contrast, if I kick a car, then I may *damage* the car, and I may harm or wrong *the owner*. But there is no sense in which I harm or wrong *the car*. We can mark this distinction by saying that the moral circle includes cats but not cars; that cats matter for their own sakes, whereas cars matter merely for our sakes.

When philosophers ask about the moral circle, we have a variety of moral questions in mind. For example, which kinds of beings can be members of the moral circle, and why? Does every member of the moral circle matter equally, or do some (say, members of our nation, generation, and species) matter more than others (say, members of other nations, generations, and species)? And does every member of the moral circle matter in the same kind of way, or do some (say, beings with the capacities for advanced language and reason) matter in different kinds of ways than others (say, beings without such capacities)?

The LaMDA case illustrates our confusion about which beings belong in the moral circle. For example, does the moral circle include all *sentient beings*, that is, beings who can experience happiness and suffering? Does it include all *agents*, that is, beings who can set and pursue their own goals? Does it include all *living beings*, that is, beings who can survive or reproduce? Does it include other kinds of beings? Moreover, what follows for LaMDA? Can a being made out of silicon be sentient, agential, alive, or otherwise morally significant? And if we disagree or feel uncertain about that, how should we treat it—or, perhaps, *them*?

Meanwhile, the Happy case illustrates our confusion about what we owe beings who belong in the moral circle. All parties involved in this case, ranging from the Nonhuman Rights Project to the Bronx Zoo to the New York Court of Appeals, agree that elephants morally matter for their own sakes. Yet they disagree about whether elephants have the same basic moral, legal, and political rights that humans have. Specifically, does Happy have a right to bodily liberty, such that her state of captivity at the Bronx Zoo constitutes an unjust detention? And if we disagree or feel uncertain about that, how should we treat her?

These questions are important. For all we know, a *vast number* and *wide range* of beings belong in the moral circle, and increasingly, our actions and policies determine whether they can exist at all and what kinds of lives they can have if they do. Moreover, the vast majority of these beings are radically different from us. The invertebrate population is much larger than the vertebrate population, and in the future, the digital population could be much larger than the physical population. Since the populations that we impact the most are the ones that we understand the least, we need to ask what we might owe them.

These questions are also difficult. The basic issue is that one kind of being—humanity—has a lot of influence over other kinds of beings, and we need to decide how to treat them without always knowing whether they matter or what we owe them if they do. Inevitably, bias and ignorance will limit our ability to do this work well. How can we exert our influence responsibly when we have so much incentive to place our thumbs on the scales for humanity? And even if we aspire to be impartial, how can we exert our influence responsibly when we have no idea what it might be like to be other kinds of beings?

The importance and difficulty of these issues raises two themes that will be important throughout this book. The first is that thinking about the moral circle requires embracing disagreement and uncertainty. On the one hand, clarifying what we owe nonhumans requires solving some of the hardest problems in ethics and science. This will not happen overnight. On the other hand, human practices are already affecting quadrillions of nonhumans every year. We need to find ways to assess these practices now, rather than wait for an imagined ethical and scientific clarity that will take a long time to arrive, if it ever does.

The second theme is that thinking about the moral circle requires embracing both our strengths and our limitations as humans. On the one hand, we have the ability to improve our social, legal, and political systems over time, achieving and sustaining higher levels of support for vulnerable populations than previously seemed possible. On the other hand, there will always be limits to what we can achieve and sustain, and we will never be able to treat everyone as they deserve. We have no choice but to embrace both of these realities, seeking to make as much progress as possible without allowing our reach to exceed our grasp.

Of course, this predicament could change in the future. For example, if AI development continues at its current pace, then AI cognition could one day rival or surpass human cognition, and AI systems could one day be able to make moral decisions with us—or even for us. If this future comes to pass, then how we draw the moral circle during our time in power might affect how AI systems draw the moral circle during their time in power. And even if not, thinking about how we would feel in such a future can still be a useful exercise, since it can help us to draw the moral circle in a more equitable manner now.

At least in the West, the history of thinking about the moral circle has been one of *moral circle expansion*. As a general trend, we started with highly exclusionary views, because we overestimated what it takes to matter and underestimated who has what it takes. We then corrected these mistakes gradually over time. For instance, many experts previously believed that only rational beings matter, but many experts now believe that all sentient beings matter. Similarly, many experts previously believed that only vertebrates can be sentient, but many experts now believe that many invertebrates can be sentient too.

Meanwhile, humanity is currently expanding our moral circle

in another, more practical way as well. Our impact on beings—both present and future—is intensifying due to advances in technology, increases in industrial activity, and resulting global catastrophes like climate change. We are also increasingly *creating* beings, including both animals and AI systems, who at least might be morally significant. Humanity is thus potentially impacting many more potentially morally significant beings than we did in the past. We need new moral frameworks to deal with this predicament.

This book will argue that humanity should expand our moral circle much more than we have thus far, for two reasons. First, a vast number of beings might matter, including insects and, eventually, AI systems, and so we should extend at least some moral consideration to these beings. Second, our actions might be affecting a vast number of beings, including beings who are far away in space and time, and so we should extend at least some moral consideration to these beings, too. The upshot is that we should extend at least some moral consideration to *sextillions* of beings, including and especially *future nonhumans*.

Chapters 1 and 2 examine current debates about moral status and moral theory. Living an ethical life requires thinking about which beings matter and why, as well as about what we owe to everyone who matters and why. I consider a wide range of views about both topics, and I argue for several principles that I take to be plausible and widely accepted in contemporary ethics. For example, I argue that if a being is *both* sentient *and* agential, then they belong in the moral circle, and we have a responsibility to avoid harming them unnecessarily, and to reduce or repair this harm where possible, if we do harm them.

In Chapters 3 and 4, I argue that we should extend moral consideration to many beings. I base this argument on a moral premise and an empirical premise. The moral premise is that if a being has a non-negligible chance of being sentient, agential, or otherwise significant, then we should extend them at least *some* moral consideration in the spirit of caution and humility. And the empirical premise is that many beings have a non-negligible chance of being sentient, agential, or otherwise significant, including insects and, eventually, AI systems. Thus, we should extend *many* beings at least *some* moral consideration.

In Chapters 5 and 6, I argue that these duties apply to many beings *across space and time* as well. I base this argument on similar moral and empirical premises. Morally, if our actions have a non-negligible chance of harming someone, then we should consider that risk. And empirically, *many* actions have a non-negligible chance of harming *many* beings across space and time. For instance, factory farming and AI development impose risks and harms on many individuals, both directly and indirectly, and we owe it to current *and* future humans *and* nonhumans to mitigate these risks and harms where possible.

In the final two chapters, I consider the general implications of my arguments. We should reject human exceptionalism, which holds that humans always take priority over nonhumans, individually and collectively. In its place, we should accept a more balanced view about our moral priorities. This view holds that we should continue to prioritize humans for now, for both moral and practical reasons. But we should also prioritize humans much less, and nonhumans much more, than we do. And as our ability to support nonhumans increases over time, our responsibility to support them will increase as well.

As these chapter summaries suggest, my arguments in this book will have two general features that are worth making explicit now. First, my arguments will be *pluralistic*. We have already seen that there is still a lot of disagreement and uncertainty about both ethics and science. Some experts think that only sentient beings matter and that only animals are sentient, whereas others think that some non-sentient beings matter and that some non-animals are sentient. To the extent possible, I will restrict myself to general assumptions that many experts can endorse despite disagreement or uncertainty about these issues.

Second, and relatedly, my arguments will be *probabilistic*. Our lack of clarity about the moral circle has implications for how we discuss it. Specifically, instead of simply discussing which beings matter and what we owe them, we should discuss which beings *might* matter and what we *might* owe them. By discussing higher or lower levels of confidence about particular views rather than simply accepting or rejecting them entirely, we can cultivate the virtues of caution and humility. To the extent possible, I will state my assumptions in these terms, expressing confidence in them while leaving room for doubt.

My aim in this book is thus not to defend any specific view about the moral circle. My aim is instead to argue, as clearly and concisely as possible, that *many beings might matter and we might owe them a lot*. Many people accept that sentient agents belong in the moral circle, that we have a duty to reduce and repair the harm that we cause them where possible, and that we should consider non-negligible risks when deciding what to do. I will show that these simple principles, together with our current predicament, imply that we have a duty to consider a *vast number* and *wide range* of current *and* future humans *and* nonhumans.

When the New York Court of Appeals denied that Happy has a right to liberty, they were clearly attempting to block a slippery slope. If we accept that *this* nonhuman has *this* right, must we also accept that *other* nonhumans have *other* rights? For example, must we accept that farmed animals have a right to liberty? That wild animals have a right to security? That animals, plants, or other organic beings have a right to life? And—if Google was wrong about LaMDA, or if future

AI systems might be relevantly different—that robots, chatbots, or other digital beings have a right to autonomy? If so, what will become of our societies?

The standard response to this worry is that we can block this slippery slope by drawing a line somewhere between elephants, on the one hand, and insects and AI systems, on the other hand. I will defend a different response: *We should not try to block this slippery slope at all.* Yes, if we dive down headfirst, then we risk breaking our necks. But if we take precautions—if we hire a lifeguard, inspect the slide, fold our arms, and slide feet first—then the benefits of taking the ride outweigh the risks. We should thus expand the moral circle to include everyone who might matter, provided we do so thoughtfully.

If the arguments in this book trouble you, please know that they trouble me, too. I set out to argue that insects and AI systems merit moral consideration, yet I ended up persuading myself that plants and microbes do as well. If correct, these arguments have transformative and potentially destabilizing implications. Hopefully, the fact that I feel troubled by my own arguments is a sign that I developed them in good faith, and not as a mere rationalization for a predetermined conclusion. In any case, all I can do now is invite you to be troubled alongside me, so that we can figure out what to do together.

CHAPTER 1

Moral status

In 1948, the United Nations made history by ratifying the first Universal Declaration of Human Rights. Member states codified their commitment to "the inherent dignity and . . . the equal and inalienable rights of all members of the human family" and pledged to promote "universal respect for and observance of human rights." Three years after the end of World War II and the Holocaust, a universal declaration of human rights was a major achievement. For the first time, the nations of the world came together to assert that all humans matter equally, independently of race, sex, language, religion, or other such features.

The idea of universal basic human rights and equality is now widely accepted. Of course, this is not to say that this idea is *universally* accepted. Humans still divide into groups based on perceived differences and assert that some groups matter more than others. And even where this idea is accepted in theory, it can still be difficult to apply in practice. Our legal, political, and economic systems still perpetuate disparities, producing better outcomes for some groups than others due to past and ongoing injustices. Still, the fact that so many humans now condemn such injustices is a major step in the right direction.

Suppose we accept—as, I think, we should—that all humans have universal basic rights. Should we also accept that *only* humans have such rights? Some say yes on principled grounds: They see humanity as the *source* of universal basic rights, either because they see humanity as valuable in itself or, at least, because they see humanity as the source of *other* valuable features, such as advanced language and reason. Others say yes on practical grounds: Since justice for humanity is still a work in progress, they believe that we should achieve justice for our own species before we start pursuing justice for any other species.

Yet humanity is only one species among many. And while there are many differences between humans and other animals, there are many similarities too. Many species have basic capacities for perception, learning, memory, anticipation, self-awareness, social awareness, communication, problem solving, emotionality, social bonds, and other features that bring meaning and value to our lives. If these features can bring meaning and value to nonhuman lives as well, then perhaps a commitment to "universal respect for and observance of human rights" is only a small, albeit important, step on the path toward justice.

You, Carmen, and Dara are housemates. You live in a three-bedroom house and spend a lot of time together. You share meals, you watch movies, and you talk about your lives. As in all close relationships, you sometimes encounter minor issues, but you navigate them well. For example, Carmen tends to wake up late, Dara tends to go to sleep early, you tend to nap during the afternoons, and all three of you tend to be sensitive to sound. So, you all make a point to keep the sound down where possible, though you still have music playing off and on throughout most days and the occasional wild house party at night.

One day, you all decide to take a genetic test, and Carmen makes a shocking discovery. She learns that she is not, in fact, a member of the species *Homo sapiens*. Instead, she belongs to the closely related species *Homo neanderthalensis*. Experts had assumed that this hominid species was extinct, but apparently a small population survived and evolved in all the same ways that humans did. Carmen is still the same person who she always was, and you all still have the same relationships that you always had. But you have different genetic sequences, enough to make you members of different species.

Meanwhile, Dara makes an even more shocking discovery. She learns that her body is not, in fact, made out of carbon-based cells. Instead, her body is made out of functionally identical silicon-based chips, like a robot from *Westworld*. Experts had assumed that this technology was still decades away, but it turns out that development is nearly complete and a small population already exists in beta mode. As with Carmen, Dara is still the same person she always was, and you still have the same relationships that you always had. But your bodies have different material substrates, and they also have different origins.

This case, inspired by a similar case from Dale Jamieson, raises a moral question: How, if at all, do these revelations affect your *moral relationships* with Carmen and Dara? From your perspective, a lot has stayed the same: Carmen and Dara seem to have the same kinds of thoughts, feelings, projects, and relationships as before. But a lot has changed, too: Neither of your roommates is a member of the same species as you! So, do you still have duties to them or not? For example, do you still owe it to them to keep the sound down, or do you now have the right to play loud, disruptive music as much as you like?

This question concerns which beings have *moral standing*, or in other words, which beings are part of the moral circle. When a being has moral standing, they have a specific kind of intrinsic value: They matter for *their own* sakes, and we owe it to *them* to treat them well. In contrast, when a being lacks moral standing, it lacks this kind of value. It might matter for *our* sakes, and we might owe it to *ourselves* to treat it well. But it does not matter for *its own sake*, and nobody owes *it* anything at all. When we ask about the basis of moral standing, we are asking what it takes to have this kind of intrinsic value.

When you assumed that Carmen and Dara were human, you assumed that they had moral standing, that they were part of the moral circle. Yes, they mattered to you, and so you owed it to yourself to treat them well. But, you assumed, they also mattered to *themselves*, and so you also owed it to *them* to treat them well. Now that you know that Carmen is a Neanderthal and Dara is a robot, do your previous assumptions still hold? In other words, is moral standing based on features that humans, Neanderthals, and robots all have? Or is it based on features that humans have but that Neanderthals and/or robots lack?

When we ask whether a being has moral standing, we are asking whether *others* can have duties to *them*, not whether *they* can have duties to *others*. Philosophers use the terms *moral patient* and *moral agent* to mark this distinction: You count as a moral patient if *others* can have duties to *you*, and you count as a moral agent if *you* can have duties to *others*. When you assumed that Carmen and Dara were human, you assumed that they were moral patients *and* moral agents. You might now be questioning both parts of that assumption. But the present question is whether your roommates are moral patients.

Similarly, when we ask whether a being has moral standing, we are not asking whether moral agents have any *particular* duties to them. We can have different duties to different moral patients, and we can also have stronger duties to some moral patients than to others. When you assumed that Carmen and Dara were carbon-based humans, you assumed that you all had the same kinds of duties to each other. You might now be questioning that assumption as well. But the present question is whether you have duties to your roommates in general, not whether you have the same kinds of duties to them as you previously assumed.

So, what does it take to have moral standing? While philosophers disagree about this issue, many agree that we should at least accept what I will call *the welfare principle*. Roughly speaking, the capacity for welfare is the capacity to be *benefited* (that is, made *better off*) or *harmed* (that is, made *worse off*). According to the welfare principle, if you have the capacity to be benefited or harmed, then you matter for your own sake, and moral agents can have duties to you. We can unpack this idea in a moment, but we all have an intuitive grip on it. Again, kicking a car might *damage* it. But kicking a cat *harms* and *wrongs* them.

The welfare principle implies that we should reject *rationalism*, which holds that only rational beings have moral standing. We all lack the capacity for rationality early in life, many of us lose this capacity late in life, and many of us never develop it at all. Yet we still have moral standing in these moments because things can still be good or bad for us. Yes, infants might not have duties to *us*, and we might not have the same kinds of duties to them that we have to each other. But we still have duties to infants that reflect their own interests, needs, and vulnerabilities, including a duty to avoid harming them unnecessarily.

This principle also implies that we should reject *speciesism*, which holds that only humans can have moral standing. In the same way that things can be good or bad for non-rational beings, things can also be good or bad for nonhumans. As Peter Singer famously argues, you owe it to a mouse to avoid kicking them unnecessarily at least partly because being kicked is *bad for them*. Similarly, you owe it to your roommate Carmen to keep the music down in the morning at least partly because, recent revelations about her species membership notwithstanding, hearing loud music in the morning is *bad for her*.

Finally, this principle also implies that we should at least raise questions about *substratism*, which holds that only *carbon-based beings* can have moral standing. The question here is: If a being is made out of, say, silicon-based chips, can things still be good or bad for that being? This question is difficult to answer, since it raises many further questions about which kinds of features are necessary for welfare and which kinds of beings can have these features. But at least in principle, you should accept that *if* your roommate Dara has the capacity for welfare as a robot, *then* you owe it to her to keep the music down at night.

What, then, does it take to have the capacity for welfare? Again, philosophers disagree about this issue, but we can consider three common answers. First, *the sentience view* asserts that sentient beings are welfare subjects. You count as sentient when you can experience positive states like pleasure or happiness or negative states like pain or suffering. This view holds that sentient beings are welfare subjects because things are good for you when they bring you pleasure and happiness and bad for you when they bring you pain and suffering. Why is kicking a cat or mouse bad for them? At least partly because it causes them to suffer.

The sentience view implies that Carmen has the capacity for welfare because Neanderthals can clearly experience happiness and suffering. What about Dara? The answer might depend on whether sentience requires *consciousness*, or the capacity for *subjective awareness*. Dara can clearly detect positive and negative stimuli. If she touches a burning stove, her hand will send a signal to her brain, causing her to jerk her hand away and yell "ouch!" But does it *feel like anything* for Dara to be detecting these stimuli? And is this kind of subjective awareness necessary for Dara to count as sentient in the relevant sense?

Another common view is *the agency view*, which asserts that agents are welfare subjects. You count as an agent when you can set and pursue goals in a self-directed manner. This can happen in different ways; for instance, many animals engage in planning and problem solving, but only some engage in moral deliberation. Regardless, this view holds that agents are welfare subjects because things are good for you when they support you in achieving your goals and bad for you otherwise. Why is kicking a cat or mouse bad for them? At least partly because it interferes with their ability to set, pursue, and achieve their goals.

Like the sentience view, the agency view implies that Carmen has the capacity for welfare, since Neanderthals can clearly set and pursue their own goals. What about Dara? As with sentience, the answer might depend on whether agency requires consciousness. Dara can definitely make decisions. She has brain states that function like beliefs and desires as well as brain processes that function like planning and problem solving. But once again, does it *feel like anything* for Dara to be making these decisions? And is this kind of subjective awareness necessary for Dara to count as agential in the relevant sense?

Finally, consider *the life view*, which asserts that living beings are welfare subjects. You count as alive when you can perform basic functions that support survival and reproduction for beings of your kind. This can happen in different ways too; for instance, animals breathe oxygen and plants breathe carbon dioxide. Regardless, this view holds that living beings are welfare subjects because things are good for you when they support you in performing your species-specific life functions and bad for you otherwise. Why is kicking a cat or mouse bad for them? At least partly because it interferes with their biological flourishing.

Like the other views, the life view implies that Carmen has the capacity for welfare, since all animals are clearly part of the tree of life. What about Dara? That depends on whether life requires a particular kind of substrate and origin. Dara can clearly perform basic life functions; she eats, drinks, pees, and poops. Apparently she can even reproduce if she likes, albeit only with other robots. The only difference is that Dara is made out of silicon, not carbon, and is a product of science, not evolution. Are a carbon-based substrate and an evolutionary origin necessary for Dara to count as alive in the relevant sense?

The next time you have dinner with Carmen and Dara, you feel unsettled. In some ways, it feels like everything has changed. Carmen is a Neanderthal! Dara is a robot! But in other ways, it feels like nothing has changed. When Carmen tells you that she got a promotion, you want to congratulate her. When Dara tells you that she got dumped, you want to console her. And these empathetic responses are more than mere habit, more than mere practice for other interactions, more than merely a plea for reciprocated kindness: You have these responses at least partly because *your roommates still seem to have lives that matter to them.*

Is this perception accurate? With Carmen, the answer is clearly yes. Given all the similarities between humans and Neanderthals, you can be confident that her life matters to her, and that *this* should matter to you. But with Dara, the answer is less clear. Like you, Dara can detect positive and negative stimuli, set and pursue goals, and perform basic life functions. But unlike you, her body is a product of silicon and science. You also feel unsure whether it feels like anything to be her. *Does* it feel like anything to be her? And does she *need* to be conscious, or a product of carbon and evolution, to be morally significant?

Since it will be hard for me to settle all these issues here (indeed, this difficulty is part of what will drive my arguments later on), I will instead simply make two assumptions for now, both of which are acceptable from a wide range of perspectives. First, consciousness, sentience, and agency are very likely *jointly sufficient* for welfare and moral standing. If Dara can consciously experience happiness and suffering and consciously set and pursue her own goals, then she very likely has a life that matters to her and deserves moral consideration for her own sake, notwithstanding her substrate and origin.

Second, there is at least a chance that consciousness, sentience, and agency are *not all necessary* for welfare or moral standing. I will discuss this assumption in more detail later on. But for now, the basic idea is that we should allow for the possibility that our moral views are mistaken. For example, even if you feel confident that Dara needs to be conscious in order to have a life that matters to her and in order to deserve moral consideration for her own sake, you should take there to be at least a chance that your confidence is misplaced—at least a chance that you can have moral duties to Dara even if she lacks this capacity.

YOU RUN AN ANIMAL rescue center, and you have the power to support a wide range of animals. But there are always more animals in need than you have the capacity to support, and unfortunately today is no exception. You arrive at work to find both an elephant and an ant with life-threatening injuries. You determine that you can treat one or the other but not both, and that they require the same amount of time and money to treat. The only possible basis for making a decision is: Which animal has more to lose if they die? Are the stakes equal for both animals, or are they higher for, say, the elephant than for the ant?

As you ponder this decision, your team enters your office with an update. There are now *ten* elephants and *ten million* ants with life-threatening injuries. Fortunately, due to your innovative rehabilitation techniques, you have the power to treat either all ten elephants or all ten million ants at the same time. But the situation is otherwise the same, and so the only possible basis for making a decision is, once again: Which animals have more to lose if they die? Even if the stakes are higher for *a single* elephant than for *a single* ant, is it possible that the stakes are higher for *ten million* ants than for *ten* elephants?

As you ponder this decision, your team enters your office with a final update. There are now *twenty* elephants and *twenty million* ants, half of whom are made out of carbon and half of whom are made out of silicon. You can treat *either* the carbon-based elephants *or* the carbon-based ants *or* the silicon-based elephants *or* the silicon-based ants, but no more than that. The situation is otherwise the same, and so the question is, once again: Which animals have more to lose if they die? Do silicon-based animals have a stake in their lives at all, and if so, how do the stakes for carbon-based animals and silicon-based animals compare?

These questions are about the *moral weight* of lives. Do some moral patients have more at stake in life than others, and do some lives carry more weight than others as a result? Note that this question concerns only the *intrinsic* value of a life for its subject. We might also need to consider many other factors when deciding how to treat someone, such as: What kind of relationship do we have with them? What would it take for us to help them? And what would happen as a result of our helping them? We can consider these extrinsic factors in the next chapter. For now, we can focus on the intrinsic value of a life for its subject.

In his influential book *Animal Liberation*, Peter Singer argued for *the principle of equal consideration of interests*. According to this principle, all equal interests merit equal consideration, no matter whose interests they happen to be. If you have the capacity for welfare, then you also have *interests* regarding your welfare; for example, if you can experience happiness and suffering, then you have an interest in feeling happiness and avoiding suffering. And when two interests are equally strong, they have equal intrinsic value, even if one of them resides in, say, an elephant and the other resides in, say, an ant.

However, Singer also noted that equal consideration of interests is compatible with differential treatment. For example, it might be that both the elephant and the ant have an interest in avoiding suffering, but that the elephant has a *stronger* interest in avoiding suffering, because the elephant is capable of experiencing more intense suffering for longer periods of time. If so, then it might be that the elephant has more at stake in life than the ant—not because an interest matters more when it resides in an elephant (that would be speciesist!), but rather because the elephant has a stronger interest in avoiding suffering.

Should we accept that some moral patients have more at stake in life than others in this way? According to *the equal weight view*, the answer is no. We might feel tempted to say that the best possible elephant life is *better* than the best possible ant life, but that would be a mistake. Each species has a distinct form of life, and each life can be assessed only by species-specific standards. Thus, the best possible elephant life might be better *for the elephant* than the best possible ant life, and the best possible ant life might be better *for the ant* than the best possible elephant life. But neither kind of life is better as a general matter.

In contrast, according to *the unequal weight view*, some moral patients really do have more at stake in life than others. For example, since the elephant has a more complex brain than the ant, they can experience more happiness, suffering, and other such welfare states at any given time. Since the elephant has a longer lifespan than the ant, they can also experience more happiness, suffering, and other such welfare states over time. And since the elephant has a higher capacity for happiness, suffering, and other such welfare states than the ant, the best possible elephant life is, in fact, better than the best possible ant life.

While the unequal weight view might seem straightforward, a lot depends on the details. For example, how exactly are welfare capacities related to brain complexity? We might be tempted to think that if one animal has ten times as many neurons as another, then the first animal can experience ten times as much welfare at any given time. But even if our welfare capacities *generally* track our neuron counts, they might not do so in a straightforward way. For example, it might be that ten times as many neurons corresponds to *five* times as much welfare, *twenty* times as much welfare, or many other possibilities.

Similarly, how exactly is our welfare capacity related to our lifespan? We might be tempted to think that if one animal can live ten times as many years as another, then the first animal can experience ten times as much welfare over time. But that might be a mistake too. Some brains have faster "clock speeds" than others, which means that they can perform more basic operations in a given unit of time. Can brains with faster clock speeds experience more welfare in a given unit of time? If so, then it might be that our lifespan, like our neuron count, is at best an imperfect proxy for our welfare capacity.

For these and other reasons, the principle of equal consideration of interests is compatible with many views about how to compare the moral weight of lives. As your team debates whether to save the elephant or the ant, you might disagree about this issue: Some of you might hold that the stakes are equal for the elephant and the ant, others might hold that the stakes are *a bit* higher for the elephant, and others might hold that the stakes are *a lot* higher for the elephant. But as long as you agree that equal interests merit equal consideration, nobody is discriminating against either animal on the basis of their species membership alone.

Of course, matters become more complex when your team needs to compare the stakes for entire *populations*. Consider a simple example involving only elephants: If you need to choose between treating an elephant with a migraine and treating an elephant with a minor headache, then you should clearly treat the elephant with the migraine, since they clearly have more at stake in this situation. But what if you need to choose between treating *ten* elephants with migraines and treating *ten million* elephants with minor headaches? Can a large number of small harms carry more weight than a small number of large ones overall?

Similarly, suppose you accept the unequal weight view, along with the idea that an elephant has a higher welfare capacity than an ant. In this case, you might accept that *if* you need to choose between saving an elephant and saving an ant, *then* you should save the elephant, since that will prevent more suffering at present and allow for more happiness in the future. But what if you need to choose between saving *ten* elephants and saving *ten million* ants? Even if a large animal carries more weight than a small animal, can a large number of small animals carry more weight than a small number of large ones overall?

According to *the non-aggregation view*, the intrinsic value of welfare is not combinable across individuals in this kind of way. If an individual migraine is worse than an individual minor headache, then an individual migraine is worse than *any number of* minor headaches. Similarly, if an individual elephant carries more weight than an individual ant, then an individual elephant carries more weight than *any number of* ants. This view implies that your team should take the elephants to have the most at stake in this situation no matter how many other, smaller animals are enduring other, lesser harms.

In contrast, according to *the aggregation view*, the intrinsic value of welfare really *is* combinable across individuals in this kind of way. A sufficiently large number of minor headaches really *can* be worse than a single migraine, even if we might not always know what that number is. Similarly, a sufficiently large number of ants really *can* carry more weight than a single elephant, even if we might not always know what that number is. This view implies that your team should follow the math: Even if you empathize with the elephants more, you should allow for the possibility that the ants have more at stake overall.

In his influential book *Reasons and Persons*, Derek Parfit showed that both of these views can have seemingly implausible implications in some cases. For example, the non-aggregation view seems to imply that a future with ten billion happy humans is worse, overall, than a future with ten *very* happy humans. That seems implausible. However, the aggregation view seems to imply that a future with ten billion happy humans is worse, overall, than a future with ten sextillion *barely* happy humans—humans who do nothing but listen to Muzak and eat potatoes all day. That seems implausible too!

Parfit famously found this implication of aggregation—that a large number of humans listening to Muzak and eating potatoes all day could be better off, overall, than a small number of humans leading complex, varied lives—so implausible that he called it *the repugnant conclusion*. Many philosophers still use this name for this idea, and while I have issues with the name, I also love a good pun. So in honor of Parfit, when I discuss the idea that a large number of small beings (say, ants) can carry more weight overall than a small number of large beings (say, elephants), I will call this idea *the rebugnant conclusion.*

Should we, like Parfit, resist this conclusion? Without attempting to answer this question here, I will offer two notes of caution. First, ethics is a marathon, not a sprint. Since all moral theories have seemingly implausible implications, we need to compare moral theories holistically before we can select one. Second, the risk of human bias is pervasive. Since our perspectives are shaped at least partly by our personal interests, our inability to empathize with small animals, our inability to imagine large numbers, and other such factors, we need to take our moral intuitions with a grain of salt, particularly about topics like population ethics.

Unfortunately, these problems are all amplified when we attempt to make welfare comparisons across substrates, too. When your team is deciding which animals to save, you might be more confident that the carbon-based animals have the capacity for welfare than that the silicon-based animals do. But if you were to determine that the silicon-based animals *do* have the capacity for welfare, then you would need to do more than make welfare comparisons between the elephants and the ants. You would also need to make welfare comparisons between the carbon-based animals and their silicon-based counterparts.

Intersubstrate welfare comparisons are even harder than interspecies ones. As much as your team struggles to compare the carbon-based elephants and the carbon-based ants, at least you know that they have the same material substrate and evolutionary origin. To the extent that you can compare their welfare capacities at all, these commonalities are what allow you to do it. Yet with the carbon-based animals and their silicon-based counterparts, you lack these commonalities. How can you make a principled decision about which animals to save when you have no idea how much they might have at stake in this situation?

As with the debate about moral standing, I will not attempt to resolve these debates about moral weight here. Instead, I will simply assume that the principle of equal consideration of interests is very likely correct. When your team is deciding which animals to treat, you might not be sure which animals have interests or how strong their interests are. You might also need to consider other factors, such as your histories with these animals, your capacity to help them, and the effects of helping them. But to the extent that these animals *do* have equally strong interests, you can be confident that their interests are equally intrinsically valuable.

Meanwhile, I will remain neutral about whether large animals carry more weight than small ones, as well as about whether large numbers of small animals can carry more weight than small numbers of large ones. My goal in this book is not to resolve such issues, but rather to argue that many nonhumans belong in the moral circle and that humans might not *always* take priority. As we will see, we can defend these ideas no matter what we think about the moral weights of lives, since soon enough, we might share the world with much larger numbers of much smaller nonhumans *and* with much smaller numbers of much larger ones.

If the United Nations were to release a *truly* universal declaration of rights, what would this declaration be like? Would it extend rights to all sentient beings, all agents, all living beings, or a different category entirely? Also, would this declaration extend equal rights to all of these beings, or would it extend stronger rights to some of them (say, individual elephants) than to others (say, individual ants)? And either way, what would the implications be in a global community that includes large numbers of small beings and small numbers of large ones? Would the United Nations embrace the rebugnant conclusion or not?

It might seem pointless to ask these questions. When most legal and political authorities still resist the idea of rights for individual chimpanzees and elephants, what hope can there be for large numbers of insects or AI systems? We might reasonably wonder whether these beings are morally significant at all. And even if we set that issue aside, we might also reasonably wonder whether we have the knowledge, power, and political will necessary to achieve and sustain even minimal levels of support for such large and diverse populations, given our apparent inability or unwillingness to do so even for members of our own species.

However, the idea of nonhuman rights might not be as far-fetched as it appears. Many governments now recognize that many animals merit moral, legal, and political protection for their own sakes. For example, France, Quebec, Colombia, Mexico City, and other governments now recognize many animals as "sentient beings." And the Constitutional Court of Ecuador, the Delhi High Court in India, the Islamabad High Court of Pakistan, and other legal authorities now recognize some animals as rights-holders. Perhaps these developments will pave the way for wider recognition of nonhuman rights in the future.

Before we continue, it helps to step back and take perspective. Two hundred years ago, many humans found the idea of universal human rights odd. Today, however, a higher percentage of humanity recognizes these exclusionary and hierarchical attitudes for what they are: tools of oppression, motivated by bad ethics and science, a desire for power and privilege, and a failure to imagine more inclusive and egalitarian societies. Two hundred years from now, will our successors view our attitudes about nonhumans the same way? And if the answer might be yes, then how should we start viewing these attitudes now?

CHAPTER 2

Moral theory

IN 2015, THE UNITED NATIONS launched an ambitious blueprint for the future: the Sustainable Development Goals (SDGs), a collection of 17 goals and 169 targets designed to steer the planet toward a more just and sustainable future. The SDGs include bold objectives like ending poverty and hunger; improving education and healthcare; promoting gender and racial equality; promoting clean water and energy; making cities and communities more resilient and sustainable; protecting forests and oceans from human-caused climate change; and ensuring peace, prosperity, and partnership for all nations in the process.

The SDGs reflect the increasingly common view that we have moral duties that extend beyond our spatial and temporal borders. For example, a 2013 report commissioned by the UN asserts that higher-income countries have a duty to assist lower-income countries on the grounds that higher-income countries "have huge potential" to impact lower-income countries. And a 2021 UN initiative applies this analysis to future generations too, noting that "decisions made today will shape the course of the planet for centuries" and that future generations "will have to live with the consequences of our action and inaction."

Of course, as with declarations involving human rights, declarations involving international and intergenerational duties reflect our ideals more than our practices. For example, the United States is the wealthiest nation in the world, yet it spends less than 1% of its annual gross domestic product (GDP) on foreign aid, and it spends a similarly paltry amount on efforts to make its own infrastructure more resilient and sustainable for the future. And while an increasing number of nations recognize future generations in their constitutions, this recognition has not yet led to significant policy change.

Moreover, while the SDGs at least *aspire* to treat members of other nations and generations with respect and compassion, they involve no such aspirations for members of other species. The 2030 Agenda for Sustainable Development makes several references to the value of biodiversity, calling for a world "in which humanity lives in harmony with nature and in which wildlife and other living species are protected." Yet none of the 17 goals or 169 targets references the health, welfare, or rights of individual animals at all. How can our policies be just when we ignore more than 99% of affected individuals?

David manages a factory at the edge of town. The factory sits on a plot of land near a park, which includes a grassland area, a woodland area, and a lake. David takes a walk in the park every day, and he regularly notices animals. There are birds in the trees, squirrels on the grass, rabbits in the woods, fishes in the lake, and more. Over time, David has even started to recognize many of these animals as individuals; for instance, he knows which rabbits are comfortable around humans and which are not. He enjoys seeing and hearing the animals, and he also enjoys sharing space with them and thinking of himself as part of nature.

The factory that David manages produces chemical waste, and he has two options for disposing of the waste. First, he can use a chemical and physical treatment process to neutralize and remove hazardous materials from the waste, and he can then safely dump the treated waste in the lake. Second, he can skip this process and simply dump the untreated waste in the lake. The first option would be safer for local wildlife, but it would add an extra step to the process and an extra line to the budget. David can afford to spend the time and money, but he prefers to save where he can, and so he selects the second option.

Unfortunately, this waste-disposal process gradually transforms the environment, and soon enough, the animals start to get sick. David once again has two options. First, he can stop dumping untreated waste in the lake, and he can also apply a treatment process to the lake to mitigate the damage caused thus far. Second, he can continue dumping untreated waste in the lake and simply let the animals get sick and die. As before, the first option would be safer for the local wildlife but more expensive for David. And as before, David can afford this cost, but he prefers to save, so he selects the second option.

Here is the question that interests me: Is David morally permitted to treat the local wildlife this way? Does he have a right to simply expose the animals to untreated waste? Moreover, if he does expose them to untreated waste, does he have a right to simply allow them to suffer and die at that point? Finally, if he does have a duty to avoid or mitigate this harm, *why* does he have it? Does this duty derive from a general duty to help or avoid harming vulnerable beings? Or does it instead derive from other, more specific features of this situation, such as how David feels about or interacts with these particular animals?

These questions concern what moral agents owe to moral patients. As noted in the previous chapter, you count as a moral agent if *you* can have moral duties to *others*, and you count as a moral patient if *others* can have moral duties to *you*. We can assume that David counts as a moral agent, since he has the ability to rationally assess his actions, and we can also assume that the animals count as moral patients, since they have the ability to be benefited or harmed. As a result, we can assume that David can have moral duties to these animals. The question is what, if anything, David owes these animals in this situation.

When we ask what moral agents owe moral patients, we should always keep in mind the distinction between *moral theory* and *moral practice*. When we think about moral theory, we ask what makes right actions right. For example, does David have a general duty to help vulnerable others, and if so, why? In contrast, when we think about moral practice, we ask how to make good decisions in everyday life. For instance, if David *does* have a general duty to help vulnerable others, then what strategies can he use to accomplish this goal in particular contexts, despite the limitations on his time, information, and rationality?

The distinction between theory and practice is helpful in part because it would be difficult to make good decisions in everyday life through theoretical deliberation alone. Consider an analogy with science. In theory, we can assess the trajectory of a baseball by calculating its initial velocity, its launch angle, and other such factors. But in practice, we lack the time, information, and rationality necessary to perform such calculations, and so we need to rely on a combination of experience, muscle memory, and simple heuristics instead. This strategy might not be perfect, but it might be the best we can do under the circumstances.

Ethics is similar. In theory, we might be able to assess the rightness of an action by calculating the extent to which it promotes welfare, respects rights, or achieves other such goals. But in practice, we might once again lack the ability to perform such calculations, and so we might once again need to rely on other strategies for making good decisions. Throughout the book, I will suggest that the distinction between theory and practice allows us to select better moral theories *as well as* better actions and policies. It also allows us to build coalitions by identifying actions and policies that multiple moral theories can endorse.

As a starting point, I want to suggest that several moral theories can endorse what I will call *the harm reduction principle*, which occupies a middle ground between theory and practice. According to this principle, we should avoid harming others unnecessarily, and if we do harm others unnecessarily, then we should help them where possible. We should also cultivate habits, relationships, and other structures that support this work. We can unpack this idea in a moment, but we all have an intuitive grip on it. You should avoid hitting cyclists with your car, and if you do hit a cyclist, then you should help them seek medical assistance.

A few features of the harm reduction principle are worth noting right away. One is that while we might or might not *always* have a duty to help others, we *do* have a duty to help others in cases where we harmed them in foreseeable and unnecessary ways. For instance, if David sees a stranger dumping untreated waste in the lake, then the harm reduction principle is neutral about whether David should help the animals. But if *David* dumps untreated waste in the lake, then this principle implies *not only* that he was wrong to harm these animals but *also* that he would be wrong not to help them, given his responsibility for their predicament.

A second important feature of the harm reduction principle is that the duty to avoid harming others is potentially overridable in cases where harming others is *morally necessary*. For example, if David *needs* to kill a particular animal to defend himself, to defend someone else, or as an unavoidable side effect of a morally necessary activity, then he might have a right—or even a duty—to kill this animal. But no such reason applies in this case. David is not dumping untreated waste in the lake to defend himself or anyone else. And even if *managing the factory* is a morally necessary activity, *dumping untreated waste in the lake* is not.

A third important feature of the harm reduction principle is that the duty to help others is potentially overridable in cases where helping others is *morally impossible*. For example, if David has no way to help a particular animal without harming or wronging this animal or other animals as a means to this end, then he might not have a duty—or even a right—to help this animal, even if his own actions are what harmed them. But no such reason applies in this case either. David can help these animals by treating the waste, and we can stipulate that these actions would not in any way, shape, or form be morally wrong.

We can see why different philosophers accept the harm reduction principle by considering several leading moral theories. First, *consequentialism* holds that morality is primarily about *consequences*. We have a duty to do the most good possible in the world. Different consequentialists interpret this idea in different ways, but the standard interpretation is that we have a duty to maximize positive welfare in the world (by producing as much pleasure, happiness, and other such states as possible) and minimize negative welfare in the world (by producing as little pain, suffering, and other negative welfare states as possible).

Consequentialism endorses the harm reduction principle because, all else being equal, harming others makes the world worse and helping others makes the world better. In fact, consequentialism holds that we should help others where possible *whether or not* we previously harmed them, simply because they need help and we have the power to help them. Thus, consequentialism implies that David should avoid harming the animals in this case because harming them makes the world worse. It also implies that when he does harm them, he should help them because helping them makes the world better.

Second, *rights theory* holds that morality is primarily about rights. We do not have a duty to pursue any particular goal in life, including the greatest good for the greatest number. Instead, we have a right to pursue whatever goals we like, provided that we allow others to do the same. Different rights theorists interpret this idea in different ways, but a standard interpretation is that we have only an *imperfect* duty of *beneficence*, which means that we should *sometimes* make others *better off*. However, we have a *perfect* duty of *non-maleficence*, which means we should *always* avoid making others *worse off*.

Rights theory endorses the harm reduction principle because harming others is an expression of maleficence, and helping those we have harmed is an expression of beneficence *and* non-maleficence, since it makes others better off *and* reduces the extent to which our past actions make them worse off. Thus, rights theory implies that David should avoid harming the animals in this case because harming them is a maleficent act. It also implies that when he does harm them, he should help them because, if you harm someone unnecessarily, then you should attempt to reduce or repair the damage where possible.

Third, *virtue ethics* and *care ethics* hold that morality is primarily about our character traits and relationships. We might like to think that we make all of our decisions through rational and moral deliberation, but in reality, many other factors shape our decisions too. These factors include our emotions and other psychological dispositions, and they also include our interpersonal bonds and our broader social, political, economic, and ecological circumstances. Thus, instead of focusing on how we deliberate, virtue ethics and care ethics focus on how we can cultivate virtuous characters and caring relationships.

Virtue ethics and care ethics endorse the harm reduction principle because our treatment of others, our character traits, and our societal structures are all mutually reinforcing. How we feel about each other shapes how we treat each other, which shapes how we organize society, which shapes how we treat each other, which shapes how we feel about each other. Thus, these theories imply that David should treat these animals well not only as an expression of kindness here and now, but also as an investment in a kinder future, a shared world in which our character traits and societal structures all guide us toward collective flourishing.

So, if David were to consult these moral theories, he would find that all roads lead to the same destination: He should treat the waste at the factory instead of simply dumping it in the lake. And if he *does* dump untreated waste in the lake, then he should treat the waste at that point instead of simply allowing the animals to suffer and die. Yes, David might not be sure what makes right actions right in theory. But when these moral theories all endorse the same general course of action in practice, albeit for different reasons and with different points of emphasis, he can be confident that this general course of action is correct.

Of course, David might not always have this kind of moral clarity. Recall the case in which he witnesses a stranger dumping untreated waste in the lake. In this case, different moral theories might have different implications about whether David should help the animals; consequentialism might require him to help the animals, rights theory might permit but not require him to help the animals, and virtue and care ethics might land somewhere between these extremes, depending on the details. We can ask how to approach hard cases where moral theories appear to diverge both in theory and in practice later on.

In the meantime, I will make two assumptions that I take to be plausible and widely accepted. The first is that the harm reduction principle is very likely correct. David can be confident that he should avoid harming others unnecessarily, and that if he *does* harm others unnecessarily, then he should help them where possible. He can also be confident that he should cultivate habits, relationships, and other structures that support this kind of harm reduction. This principle is compatible with every moral theory that we have discussed, and it has clear implications for how David should treat these animals.

My second assumption is that there is at least a *chance* that we should help others where possible *whether or not* we harmed them. Consequentialists think that we have this duty, many rights theorists think that we have at least an "imperfect" version of this duty, and many virtue and care ethicists see helping others as a means to, and a result of, cultivating virtuous characters and caring relationships. David should take there to be at least a chance that one of these views is correct—at least a chance that he should help the animals whether or not he harmed them, simply because they need help and he has the power to help them.

One morning, David thinks of a novel solution to his waste problem. Instead of dumping his untreated waste in a nearby *lake*, he can dump it in a nearby *river*. The river would then carry the waste across the border, where it would settle in a similar lake in a different country. The waste would still kill animals in a foreseeable and unnecessary way, but the details would be different. Instead of killing animals *here*, the waste would kill animals *in a different place*. And instead of killing animals *now*, it would kill animals *at a different time*. David likes this solution: out of sight, out of mind. So he starts dumping his waste in the river.

Later that week, David thinks of another novel solution. Instead of dumping the waste in the river, he can bury it underneath the river bed. The waste would then stay there for about twenty-five years, at which point natural forces would release it into the river, with the same results as before. Once again, the waste would still kill animals in a foreseeable and unnecessary way, but the details would be different. In this scenario, *none of the victims exist yet*. Instead, everyone who will die as a result of this waste will be born in the time between action and impact. David once again likes this solution, so he starts burying his waste.

However, the more David thinks about the issue, the more he wonders whether these changes make a moral difference. On the one hand, he has the intuition that he owes more to those nearby than to those far away, in part because he has closer relationships with those nearby, and in part because he has a greater ability to predict and control his impacts on them. On the other hand, he has the intuition that he still owes *something* to those far away, particularly when he has at least *some* relationship with them and at least *some* ability to predict and control his impacts on them. So he still feels somewhat uncertain about his new policy.

These developments raise questions about the moral significance of *distance*. The harm reduction principle holds that we should reduce and repair the unnecessary harm that we cause where possible. Does distance in space or time affect this duty? For example, does David have a stronger duty of non-maleficence to animals one mile away than to animals twenty-five miles away, or to animals in one year than to animals in twenty-five years? And how, if at all, does this duty apply to future generations? For instance, does David have a duty of non-maleficence to animals who exist at present *as well as* to animals who will exist in the future?

The principle of equal consideration of interests, which we discussed in the last chapter, has implications for the moral significance of distance. As a reminder, this principle holds that all equal interests have equal intrinsic value, no matter which beings happen to possess them. Importantly, this principle applies not only across *species* and *substrates* but also across *space* and *time*. For instance, if two elephants have equally strong interests in avoiding suffering, then I should regard their interests as having equal intrinsic value even if one of the elephants is right next to me and the other is thousands of miles away from me.

However, as we also saw in the last chapter, equal consideration can be compatible with differential treatment. And this might be true not only for *capacities-based* reasons (some beings might have more, or stronger, interests than others) but also for *relational and practical* reasons. Plausibly, I owe more to my family than I owe to yours, not because my family has more *intrinsic* value than yours, but rather because I have a closer relationship with them and a greater ability to support them day in and day out. We might think that similar considerations can apply to members of our species, substrate, nation, and generation, too.

Consider that we can have special duties in the context of special relationships. Family is a paradigmatic example because of how close families can be. To the extent that my family members share our lives with each other, we develop special bonds of care and interdependence with each other, and we take on special duties to each other as a result. We make promises to each other, and so we take on a duty to keep those promises. We help each other out, and so we take on a duty to return the favor. We hurt each other (accidentally!), and so we take on a duty to reduce and repair the damage. Entangled lives require entangled ethics.

While family might be a paradigmatic example, other, larger kinds of groups can be examples too. For example, to the extent that members of my species, nation, and generation share our lives with each other, we take on special duties to each other, too. Yes, we might not always interact directly or personally, and so we might not always owe each other the same kinds of care that many of us owe our family members. But we might still interact indirectly and impersonally, via shared social, political, and economic systems, and we might still take on at least some special duties to each other as a result of these broader entanglements.

We also need to consider practical factors when deciding what to do. For example, we generally owe more to our families not only because we have special relationships with them but also because we have a greater ability to support them. I should provide my family with food, water, and affection on a daily basis in part because I *can*. I know what my family members need and how to provide it for them, and I can achieve and sustain this level of support for them on a daily basis. In contrast, I have no such duty to *all* families, in part because I could never achieve or sustain this level of support for all families at once.

Other, larger groups can exemplify this phenomenon, too. The United States should provide all humans who live within its borders with education, healthcare, and public safety in part because it *can*. It knows how to provide these goods to this population, and the population is small enough that it can afford to do so. But at least at present, the United States has no such duty to, say, all *nonhumans* who live within its borders, all humans who live *beyond* its borders, or, especially, all nonhumans who live beyond its borders, in part because it could never achieve or sustain such a high level of support for all these populations at once.

The principle of equal consideration of interests is thus compatible with the idea that David owes more to local wildlife insofar as he has special relationships with them and a greater ability to support them. But note three caveats. First, he might sometimes have special relationships with, or a greater ability to support, distant wildlife too. Second, he can have a duty to *avoid harming* distant wildlife whether or not he has a duty to *help* them. And third, other factors can be relevant in this case too, including the stakes for everyone. We will return to these caveats in later chapters and see how they might affect our priorities.

In addition to taking distance across space, time, biology, and materiality to matter for these reasons, some philosophers take distance across time to matter for a more fundamental reason concerning the nature of harm. We commonly think that actions are *beneficial* when they make particular individuals *better off* than they would otherwise be and *harmful* when they make particular individuals *worse off* than they would otherwise be. For example, when David dumps his waste in the lake, we might think that he harms the animals by causing them to experience less happiness and more suffering than they would otherwise experience.

However, this view about the nature of harm raises a problem for intergenerational ethics. Suppose that you plan to have a child. If you save for the future, then you can expect to have a child named Alice who will have a *very* good life. If not, then you can expect to have a child named Alex who will have a *barely* good life. Intuitively, not saving for the future is *bad for your child*. But which child? Not Alice, since she exists only if you save. And not Alex, since she exists only if you *fail* to save. So, if harming someone requires making them worse off than *they* would otherwise be, then how is failing to save for the future bad for your child?

Consider two opposing answers to this question. According to what we can call *the subject-relative view*, actions are harmful only when they make particular subjects worse off than they would otherwise be. This view implies that your failure to save for the future is *not* bad for your child. Yes, you still harm your future child if you make them worse off than *they* would otherwise be, or if you make their life *not worth living at all.* But you do not harm them simply by making them worse off than *another* child would otherwise be. If they have a life worth living, how can the choices that allow them to exist be bad for them?

In contrast, according to what we can call *the subject-neutral view*, actions are harmful when they make lives worse *in general.* This view implies that your failure to save for the future *is* bad for your future child. Yes, it might not be bad for *any particular* future child. But it still counts as bad, since it still causes your future child to experience much less happiness than your future child (albeit a *different* future child) would otherwise experience. If your reckless spending has the foreseeable and avoidable consequence that your child can, at best, have a life barely worth living, then how can morality not register that impact?

In the same book that introduced the repugnant conclusion to the world, Derek Parfit examined these questions about future harms and benefits in detail, and he noted that the implications are potentially vast. After all, *many* actions and policies might change which subjects can exist in the future. Suppose that if the United States fails to save for the future, then our descendants will experience much less happiness overall. Intuitively, this policy is *bad for our descendants.* But to the extent that it creates descendants who exist only in this future and who at least have lives worth living, that intuition might or might not be correct.

Parfit called this problem *the non-identity problem* since it concerns actions and policies that change the identities of future subjects. And while Parfit focused mostly on *intraspecies* identity changes (that is, actions that create one set of humans instead of another), this problem can arise for *interspecies* identity changes too. Suppose that by polluting the lake, David creates a future lake with *beetles* who have *barely* good lives instead of a future lake with *swans* who have *very* good lives. Is polluting the lake in any way bad or wrong in light of this impact? We can call this interspecies version of the problem *the swan identity problem.*

As with the repugnant conclusion, I will not attempt to solve the non-identity problem here. Instead, I will simply reiterate that morality is a marathon, not a sprint, and that the risk of human bias is pervasive. For example, we might initially prefer the subject-relative view because we like the idea that our actions cause less harm overall. But if so, we should ask: *Why* do we like this idea? To what extent does this attitude reveal a basic truth about the nature of harm, and to what extent does it instead merely reflect a desire to evade responsibility for harm, coupled with an inability or unwillingness to empathize with distant strangers?

Like questions about moral status, these questions about moral theory might be even harder if and when we share the world with moral patients of different substrates. Suppose that David owns the land surrounding the lake, and he can either keep it or sell it. If he keeps it, the lake will house trillions of animals in the future. If not, a company will mine the land and build a server farm thousands of miles away, and the server farm will generate trillions of digital lakes with trillions of digital animals per lake in the future. All else being equal, which outcome is better for future animals? Which option should David select?

Even when we bracket other relevant factors, this case is difficult because it combines every issue that we have discussed over the past two chapters, including which beings matter, how much they matter, what we owe each other, and the moral significance of distance. David might be uncertain about all of these issues. And insofar as he can make estimates about them, he might find that no matter what he does, future animals will be better off in some respects and worse off in other respects. How can David make a principled decision about whether to keep or sell this land when faced with so much uncertainty and complexity?

We will examine the ethics of risk and uncertainty more over the next four chapters. For now, I will simply note that the harm reduction principle and the principle of equal consideration of interests have two clear implications for David in the lake and river cases that we considered earlier in this chapter. First, he should treat the waste in the factory. He can afford to treat it, and while this action might impose a *minor* burden on him, it avoids imposing a *major* burden on the animals. Second, *if* David deposits untreated waste in the lake or river, *then* he should treat it at these locations as well, to reduce and repair the harm caused by his negligence.

Beyond that, I will remain neutral about the non-identity problem here—that is, about whether we cause harm when we create future moral patients with *barely* good lives instead of *different* moral patients with *very* good lives. But I will note that David should treat his waste either way. While many of his victims might not exist yet, his action does not determine whether they come into existence but rather only determines whether they suffer and die prematurely when they do. Thus, depositing untreated waste in the lake and river is harmful and wrong according to *both* the subject-relative view *and* the subject-neutral view.

If the United Nations were to properly consider the interests of everyone affected by international policy, what would the Sustainable Development Goals (SDGs) be like? They would give much more weight to members of other species, nations, and generations, and they would support bolder international policies as a result. For instance, member states would commit to ending factory farming for the sake of humans *and* nonhumans, they would commit to building a more accommodating infrastructure for both populations, and high-income nations would commit to supporting low-income nations in doing this work.

I know that this kind of change might seem impossible. Even when we identify policies that benefit humans in the short term, the United Nations is slow to endorse them, since endorsement requires acceptance from member states with radically different priorities. For example, we all stand to benefit from energy reform, yet nations have been slow to commit to energy reform, since they disagree about which alternatives to develop and how to pay for them. And when even energy reform is hard to achieve, more radical measures like food system reform or multispecies infrastructure might seem (literally) like a bridge too far.

However, as with the idea of multispecies rights, the idea of multispecies, multinational, and multigenerational global governance is not as far-fetched as it might appear. In 2022, the United Nations Environment Assembly called for the United Nations Environment Programme to produce a report on the links between animal welfare and sustainable development. Meanwhile, member states already accept that other nations and generations merit consideration, even if they give these distant populations less weight than they should. These developments show that progress is possible, even if painstakingly slow.

Here, too, it can be useful to step back and take perspective. Our system of global governance is still in its infancy. Two hundred years ago, the idea of international legal and political frameworks that allow member states to act collectively while retaining individual autonomy would have seemed unthinkable to many. Yet such frameworks now exist, and while they remain works in progress, they remind us that innovation is possible. What we now need is further innovation—new legal and political frameworks that give proper weight to all stakeholders, no matter how nearby or far away they happen to be.

CHAPTER 3

If you might matter, we should assume you do

In 2003 David Foster Wallace took a trip to Maine to write an article for *Gourmet* magazine about the Maine Lobster Festival. But the article that he published, titled "Consider the Lobster," is less a travelogue and more a reflection about whether killing lobsters for food causes them to suffer. As Wallace observes, this question is difficult to answer: "The questions of whether and how different kinds of animals feel pain . . . turn out to be extremely complex and difficult," since "the principles by which we can infer that others experience pain . . . involve hard-core philosophy—metaphysics, epistemology, value theory."

The problem that Wallace is identifying, which I call *the moral problem of other minds*, works as follows: Suppose that sentient beings have moral standing but we have no idea which beings are sentient. How should we make decisions in these cases of uncertainty? Wallace does not try to solve this problem. Instead, he examines the evidence concerning lobster sentience, and he concludes by expressing confusion about what to do. Many scientists and philosophers are similarly confused, though not always as self-aware. The moral problem of other minds remains both extremely important and extremely neglected.

In fact, the problem is even broader than Wallace suggested, since it applies to many more beings than lobsters. Humans regularly interact with a wide range of animals, plants, and other carbon-based beings. We increasingly interact with a wide range of robots, chatbots, and other silicon-based beings. And in the future, economic and technological developments might lead us to interact with an even richer variety of nonhumans than we already do. In all cases, if we have less than *100%* confidence about whether nonhumans have the capacity to suffer, then the moral problem of other minds applies to them.

This problem is also deeper than Wallace suggested because it presents *layers* of uncertainty. Specifically, we have uncertainty both about *values* (that is, about which features suffice for moral standing) and about *facts* (that is, about which beings have these features). For instance, you might lack clarity about whether sentience, agency, or life suffices for moral standing, as well as about whether particular nonhumans count as sentient, agential, or alive. How can you make good decisions about how to treat nonhumans when you have no idea whether they morally matter for their own sakes?

A YEAR HAS PASSED since you learned that your roommate Carmen is a Neanderthal and that your roommate Dara is a robot. You still live together and your lives and relationships are the same as before, but with a lot more existential angst. Carmen and Dara feel deeply unsettled about their revealed identities. You do your best to support them, but the process will take time and you can only do so much. Otherwise, life goes on. You still spend a lot of time with each other, you still have different interests and schedules, and you still compromise and coordinate with each other when your interests and schedules conflict.

While your focus is on supporting your roommates, you still have a lot of your own processing to do, too. Your relationships with Carmen and Dara are based at least in part on the idea that you all care about each other and have duties to each other. On a personal level, you would feel differently if your roommates were not, in fact, the kinds of beings who could have cares and concerns. And on a moral level, you would feel differently if they were not, in fact, the kinds of beings who could have duties or rights. In that case, you might still compromise and coordinate with them, but not in the same spirit as before.

So, *are* Carmen and Dara the kinds of beings who can have cares, concerns, duties, and rights? You feel confident that Carmen is, since Neanderthals are clearly conscious, sentient, *and* agential. But you feel less sure about Dara. Yes, she might *functionally* resemble a carbon-based human: She can detect positive and negative stimuli, and she can set and pursue goals in a self-directed manner. But you have no idea whether it *feels like anything* for Dara to be performing these functions. You also have no idea whether it *needs* to feel like something in order for a sense of mutual affection or responsibility to be warranted.

This example concerns the ethics of *risk*. We rarely, if ever, have enough information to warrant *certainty* about what to do; instead, we need to make do with higher or lower degrees of confidence. Specifically, we regularly struggle with *moral* uncertainty (that is, uncertainty about *values*), *scientific* uncertainty (that is, uncertainty about *facts*), or, in many cases, both. This phenomenon is so pervasive that we can easily take it for granted. But anytime you find yourself thinking that an action *might be* harmful or wrong and wondering if you should take the chance or err on the side of caution, you are assessing risks.

While philosophers disagree about the ethics of risk, many agree that we should at least accept what I will call *the risk principle*. This principle holds that all non-negligible risks—that is, all risks that have a decent chance of happening—merit consideration in our decisions. Suppose that you have reason to destroy a building, but you feel unsure whether anyone is inside. You see lights on, but you take this evidence to be compatible with someone being inside *and* with no one being inside. The risk principle implies that the possibility of someone being inside at least *merits consideration* in your decision.

The risk principle applies in cases of uncertainty about moral standing, too. Suppose that you have reason to destroy a robot, but you feel unsure whether this robot has the capacity for conscious awareness. You see that the robot has advanced and integrated capacities for perception, memory, anticipation, self-awareness, social awareness, communication, and problem solving, but you take this evidence to be compatible with "someone being inside" *and* with "no one being inside." The risk principle once again implies that the possibility of "someone being inside" at least *merits consideration* in your decision.

Of course, since the risk principle holds that *non-negligible* risks merit consideration, we need to ask which risks are non-negligible. We will examine that question shortly, but for now, we can note that even unlikely risks can merit consideration. For example, J. Robert Oppenheimer, known as the father of the atomic bomb, reportedly took there to be a chance that testing the bomb would ignite the atmosphere and destroy life on Earth. Suppose that he took the probability of this result to be only about *one in ten thousand*. In this case, clearly this risk would have at least merited consideration, despite being unlikely.

Additionally, since the risk principle holds that non-negligible risks *merit consideration*, we need to ask how to consider these risks. We will examine that question shortly too, but for now, we can note that some risks are worth taking and others are not. If testing the bomb was sufficiently important and if the risk of harm was sufficiently low, then proceeding with the test might have been a sensible decision. If not, then it might have been a reckless gamble. Either way, a decent chance of igniting the atmosphere and destroying the planet should have been, at minimum, *a factor in the decision* about whether to proceed.

To see how the risk principle applies to our thinking about which beings belong in the moral circle, consider uncertainty about values. As we have seen, debates about moral standing are ongoing. Experts still disagree about whether features like consciousness, sentience, and agency are required for moral standing. And even if we feel confident that these features *are* required, we should be open to the possibility that our confidence is misplaced. Thus far, every generation has made serious moral mistakes, including about the nature of moral standing. There is at least a *chance* that our generation is similarly fallible.

Importantly, this need for humility about our moral views will likely last for a long time. No matter what we think about the nature of moral knowledge, we can all agree on this much: Moral certainty is an elusive ideal. For moral certainty to be warranted, we would need to have *fully informed and ideally coherent* moral beliefs and values. Yet we might not be able to achieve that ideal for a long time, if ever. And as long as our beliefs and values remain at least partly uninformed or incoherent, we can reasonably endorse only *higher or lower degrees of confidence* about which features are necessary for moral standing.

The risk principle implies that if a view about moral standing has a decent chance of being correct, then this possibility merits consideration. Suppose that you take there to be a 99% chance that conscious experiences and motivations are necessary for moral standing and a 1%

chance that nonconscious experiences and motivations are sufficient. Now suppose that Dara has only nonconscious experiences and motivations. In this case, it would be wrong for you to simply conclude that Dara lacks moral standing. If you think that she has a 1% chance of mattering, then you should take this possibility into account.

Now consider uncertainty about facts. As we will see, debates about consciousness, sentience, agency, and other such capacities are ongoing as well. For example, experts still debate whether a high level of cognitive complexity, a high level of cognitive centralization, and a carbon-based substrate are required for consciousness. And even if we feel confident that these features *are* required, we should be open to the possibility that our confidence is misplaced. Thus far, every generation has made serious scientific mistakes, including about consciousness. There is once again at least a *chance* that our generation is similarly fallible.

The need for humility about our scientific views will likely last for a long time, too. The science of consciousness is particularly difficult, since nobody can directly observe what, if anything, it feels like to be someone else. Yes, we can check to see whether particular beings have "lights on," but we can never check to see whether they have "anyone inside." As a result, we might not be able to achieve certainty about which beings are conscious for a long time, if ever. And as long as we remain in this predicament, we can reasonably endorse only *higher or lower degrees of confidence* that particular beings are conscious.

You can see where this is going. If a view about consciousness has a decent chance of being correct, then this possibility merits consideration too. Suppose that you take there to be a 99% chance that a complex, centralized, carbon-based brain is necessary for consciousness and a 1% chance that a complex, centralized, silicon-based brain is sufficient. In this case, you would once again be wrong to conclude that Dara lacks moral standing *even if you grant that moral standing requires consciousness*. If you think that Dara has a 1% chance of having "someone inside," then you should take that possibility into account too.

We thus have two paths to the same destination: If sufficient uncertainty is warranted *either* about which features suffice for moral standing *or* about which beings have these features, then the risk principle applies. And if sufficient uncertainty is warranted about *both* of these issues, then this principle applies all the more. For instance, if you take there to be a 1% chance that nonconscious experiences and motivations suffice for moral standing *and* that complex, centralized, silicon-based brains suffice for conscious experiences and motivations, then you have two distinct reasons to consider your robotic roommate.

At this point, one might object that there are risks involved with including too many beings in the moral circle too. After all, we might think that *many* beings have at least a *chance* of mattering. So, if we simply assume that every feature that *might* suffice for moral standing *does* suffice for moral standing and that every being that *might* have these features *does* have these features, then we might end up taking so many beings to matter that we neglect the ones who really *do* matter, for instance by prioritizing the perceived interests of digital humans and animals over the actual interests of physical humans and animals.

This objection is well-taken. As we will see, the world contains a vast number and wide range of invertebrates, plants, microbes, AI systems, and other beings that at least *might* matter. So, if we include these beings in the moral circle, then we might feel pressure to prioritize them over humans and other vertebrates, given the relative sizes of our populations. Yet in the event that these beings are *not* morally significant (as seems likely!), it would of course be bad if we were systematically sacrificing the *actual* interests of *actual* moral patients for the merely *possible* interests of merely *possible* moral patients.

However, we can respond to this objection in several related ways. First, false negatives about moral standing may be more *likely* than false positives in many cases. If the history of our treatment of nonhuman animals is any indication, our tendency toward *anthropodenial* (mistakenly seeing nonhumans as *lacking* human traits that they *have*) may be stronger than our tendency toward *anthropomorphism* (mistakenly seeing nonhumans as *having* human traits that they *lack*) in many cases, in part because we have a strong incentive to view nonhumans as objects so we can continue exploiting or exterminating them.

Second, false negatives about moral standing may be more *harmful* than false positives in many cases. As the history of our treatment of nonhuman animals (as well as fellow humans) illustrates, the harm involved when a *subject* is treated as an *object* is generally worse than the harm involved when an *object* is treated as a *subject*. Granted, when we mistakenly treat objects as subjects, we might end up prioritizing merely *perceived* subjects over *actual* subjects in some cases. But if we engage in thoughtful priority-setting, then we can expand the moral circle substantially while mitigating this kind of risk, as we will see.

Third, and relatedly, when we consider a wider range of potential moral patients, we may find that some options are *co-beneficial* for multiple populations. For example, factory farming kills many vertebrates, invertebrates, *and* plants unnecessarily (since we need to kill more plants to feed farmed animals than we need to kill to feed humans directly). It also contributes to global health and environmental threats that imperil us all. Ending factory farming would thus benefit many human and nonhuman populations at once. And if we at least *consider* these populations, we might find that other options are similarly co-beneficial.

Fourth, we might be warranted in prioritizing humans over nonhumans in many cases even if the nonhumans merit consideration too. As we saw in the last chapter, relational and practical factors might support prioritizing fellow humans to an extent in many cases. And as we will see in a moment, some principles of risk and uncertainty might have similar implications. Thus, for example, if you need to decide whether to save your sister, your Neanderthal roommate Carmen, or your robotic roommate Dara, then you might be warranted in saving your sister for multiple reasons even if your roommates merit consideration too.

Finally, however, we should allow for the possibility that we should prioritize nonhumans over humans in some cases, too. When we give appropriate weight to all relevant factors, we might find that we should still do a lot for ourselves, but we might also find that we should make at least *some* sacrifices for nonhumans, provided that we can make these sacrifices ethically, effectively, and sustainably. Suppose that your sister reports having a minor itch, Carmen reports having a headache, and Dara reports having a migraine. If you have the power to help only one of them, is it *obvious* that your sister takes priority?

YOUR WORK AT THE animal rescue center continues. You still have a wide range of animals who appear to need help, including carbon-based elephants, carbon-based ants, silicon-based elephants, and silicon-based ants. Your team is confident that the unequal weight view and the aggregation view are correct, and so they expect that if these animals all matter, then the elephants matter more at the individual level but the ants matter more at the population level. They also expect that if these animals all matter, then the carbon-based elephants and ants matter exactly as much as their silicon-based counterparts do.

However, your team is more confident that the elephants matter than that the ants do, and your team is also more confident that the carbon-based animals matter than that their silicon-based counterparts do. Specifically, your team takes there to be about a 99.99% chance that the carbon-based elephants matter, about a 50% chance that the carbon-based ants matter, about a 9.99% chance that the silicon-based elephants matter, and about a 5% chance that the silicon-based ants matter. So, you find yourselves wondering: When deciding which animals to save, should you give less weight to the animals who are less likely to matter?

As you ponder this question, you realize that your rescue center also contains *twenty quadrillion microbes*, and that if you invest in structural changes at the center, then you can help these microbes about as effectively as you can help the elephants or the ants. Now, you take the microbes to have an *extremely* low chance of mattering, and you also take them to matter an *extremely* tiny amount, if at all. Yet there are so many of them that if you give them any weight at all individually, then you need to give them a lot of weight collectively. So, you now find yourself wondering: Should you give the microbes any weight at all?

These examples raise questions about *whether and how* to consider possibilities that involve different probabilities, including extremely low probabilities. First, should we give less weight to potential harms that have a lower chance of happening, and by extension, should we give less weight to potential moral patients who have a lower chance of mattering? Second, should we give *any weight at all* to potential harms that have an *extremely low* chance of happening, and by extension, should we give *any weight at all* to potential moral patients who have an *extremely low* chance of mattering?

We can start by considering whether we should give less weight to harms that have a lower chance of happening. We can distinguish two views about this issue. First, *the precautionary view* states that we should err on the side of caution. If a particular harm *might* occur, then we should assume that it *will* occur for practical purposes. Second, *the expected weight view* states that we should calculate expected impacts. If a particular harm *might* occur, then we should multiply the *probability* of harm by the *amount* of harm, and we should assume that the resulting amount of harm will occur for practical purposes.

How do these views apply to your rescue dilemmas? The precautionary view implies that if a being *might* matter, then you should assume that they *do* matter when making decisions that affect them. Suppose that a carbon-based ant and a silicon-based ant matter equally if they matter at all. In this case, the precautionary view implies that you should treat these ants as mattering equally when deciding whom to save. Yes, the carbon-based ant is ten times more likely to matter, but this difference should not factor into your decision-making. When in doubt, you should err on the side of attributing full weight to potential moral patients.

In contrast, the expected weight view states that if a being might matter, then you should multiply the probability that they matter by how much intrinsic value they would have if they did, and you should assume that they have the resulting amount of intrinsic value when making decisions that affect them. Thus, since the carbon-based ant is ten times more likely to matter than the silicon-based ant, the expected weight view implies that you should treat the carbon-based ant as mattering ten times more. When in doubt, you should attribute more weight to potential moral patients who are more likely to matter.

Of course, you might also feel uncertain about how likely these beings are to matter, and about how much they matter if they do. For example, you might think that an elephant is *one to twenty* times as likely to matter as an ant, and that the elephant matters *one to ten million* times as much as the ant, if at all. In that case, the precautionary view implies that you should round up and treat these beings as mattering equally, whereas the expected weight view implies that you should do more math, perhaps taking the average of these ranges and treating the elephant as mattering about *fifty million* times as much as the ant.

Which view is more plausible? In general, the precautionary view is useful in cases where some options are clearly riskier than others and our goal is to avoid worst-case scenarios. Suppose that your house is burning down and you can save any two of the following three beings: Carmen, Dara, and your pet rock. In this case, you have no need to compare the probability of harm to Carmen and the probability of harm to Dara. Leaving your roommates behind is *clearly* risky for them in a way that leaving the rock behind is not for the rock, and so you should err on the side of saving your roommates.

In contrast, the expected weight view is useful in cases where all options are comparably risky and where our goal is to maximize expected value, given the evidence available. Suppose that your house is burning down and you can save either Carmen or Dara but not both. In this case, you do need to compare the probability of harm to each roommate. And while you think that leaving each roommate behind *might* be bad for them, you also think that leaving Carmen behind is *at least a bit more likely* to be bad for her, given your shared substrate. So, perhaps you should err on the side of saving Carmen rather than simply flipping a coin.

Before we accept this analysis, however, it might help to ask how we would feel if the shoe was on the other foot. Imagine a future society in which carbon-based humans and silicon-based humans share power.

Now suppose that each kind of human views their own kind as more likely to matter given the evidence available to them. Do silicon-based humans like Dara have a right to discriminate against carbon-based humans like you on the grounds that you appear less likely to matter from their perspective? If so, perhaps you have a right to discriminate against them as well. If not, perhaps you should give all humans equal weight after all.

Now consider our second main question: whether we should give *any weight at all* to possibilities with an extremely low probability of happening. We can distinguish two views here, too. First, the *no threshold view* states that all risks merit consideration. We might be able to give extremely little weight to extremely unlikely risks, but we should give *some* weight to these risks. Second, the *threshold view* states that only some risks merit consideration. We should set a probability threshold above which risks count as *non-negligible* and below which they count as *negligible*, and we should give weight to all and only non-negligible risks.

How do these views apply to your rescue dilemmas? The no threshold view implies that a being merits moral consideration if they have *any chance at all* of mattering. For example, even if a microbe has only a *one in a trillion* chance of mattering, and even if they matter only *one trillionth* as much as an elephant, if at all, you should still give them at least *some* weight. Yes, you might be warranted in giving the microbe only a *quintillionth* as much weight as the elephant, which is equivalent to a rounding error for most purposes. Still, the microbe carries a non-zero amount of weight, and that might matter for some purposes.

In contrast, the threshold view implies that a being merits moral consideration if and only if they have a *non-negligible* chance of mattering. So, if a microbe has a one in a trillion chance of mattering, then a lot depends on whether a one in a trillion risk merits consideration. If this level of risk is non-negligible, then the threshold view has the same implications as the no threshold view: You should give the microbe at least *some* weight, even if only very little. But if this level of risk is negligible, then the threshold view has more exclusionary implications: You should give the microbe *no weight at all* in your decisions.

So, where *should* we draw the line between negligible and non-negligible risks, according to the threshold view? In the literature on this topic, philosophers typically draw the line somewhere between *one in ten thousand* and *one in ten quadrillion*. My own view is that lower thresholds are more plausible than higher ones; if pushing a button had a one in a trillion risk of destroying the planet, clearly this risk would at least merit consideration! Nevertheless, I will assume a one in ten thousand threshold here, so that I can be as generous as possible to people who might be skeptical about my arguments in this book.

Which view is more plausible? In general, the case for the no threshold view is that it takes a principled stance by holding that all potential moral patients merit moral consideration. Yes, your team might be warranted in prioritizing an elephant over a microbe for many reasons; for example, the elephant might be more likely to matter, they might be likely to matter more, you might have a closer relationship with them, you might have a greater ability to treat them well. But if the microbe has at least *some* chance of mattering, then they merit at least *some* consideration in decisions that affect them, even if only very little.

In contrast, the case for the threshold view is that it prevents extremely unlikely possibilities from dominating in our decision-making. If we take beings like microbes to matter at all, then we might find that their significance is like gravity: undetectable at small scales but inescapable at large scales. After all, even if we give much more weight to an elephant than to a microbe, we might still need to give more weight to the *total population* of microbes than to the *total population* of elephants, depending on their sizes and our views about aggregation. If we find this idea implausible, the threshold view might allow us to avoid it.

Before we accept this analysis, however, it might once again help to ask how we would feel if the shoe were on the other foot. Suppose that a godlike being wants to destroy the universe for fun. They think that humans and other earthly beings have an extremely low chance of mattering at all, but they also think that there are so many of us that if we do matter, then the risks of destroying the universe likely outweigh the benefits. Does the godlike being have a right to neglect our potential significance and destroy the universe? If so, perhaps we can similarly neglect microbes. If not, perhaps we should consider them.

The question of whether and how much to consider extremely unlikely possibilities crops up in a wide range of contexts. To adapt an example from Eliezer Yudkowsky, suppose that a mugger tells you that unless you give him five dollars right now, he plans to kill ten trillion mammals, birds, reptiles, amphibians, and fishes. He also explains how he plans to accomplish this goal, and while you take him to be *extremely* unlikely to be telling you the truth, his story is compelling enough that you take the probability to be *at least* one in ten trillion. The question, then, is: Is it rational for you to give the mugger your money?

Yudkowsky calls his version of this case *Pascal's mugging*, in honor of mathematician Blaise Pascal, who explored similar cases involving infinities centuries ago. Our discussion in this section is about what happens when Pascal's mugging meets the rebugnant conclusion. Can a sufficiently large number of very small beings (say, microbes) carry more weight than a sufficiently small number of very large beings (say, elephants) overall, even if the very small beings are extremely unlikely to matter at all and even if they matter extremely little, if at all? In honor of its predecessors, we can call this possibility *Pascal's bugging*.

Without attempting to resolve these debates, I will instead make a general assumption that I take to be compatible with all of the views discussed here. If a being has at least a *one in ten thousand* chance of mattering, given the evidence, then they merit at least *some* consideration, even if only a tiny amount. We can agree about this much even if we disagree about the details, for instance even if we hold that the risk threshold should be lower than one in ten thousand (perhaps even zero!), and even if we hold that all potential moral patients merit equal consideration rather than varying amounts of consideration.

This conservative assumption is more than enough to warrant *substantial* moral circle expansion. As we will see in the next chapter, a vast number of beings have at least a one in ten thousand chance of being conscious, sentient, and agential, given the evidence. And if more permissive theories of moral standing have at least a one in ten thousand chance of being correct, then a vast number of beings merit consideration for that reason, too. Either way, the implication is clear. Given ongoing uncertainty about both facts and values, a vast number of beings merit at least *some* consideration, even if only a tiny amount.

NEARLY TWENTY YEARS AFTER David Foster Wallace urged us to consider the lobster, philosopher Jonathan Birch and colleagues published a detailed review of the evidence of sentience in cephalopod mollusks and decapod crustaceans, including lobsters. This review examines a number of recent scientific studies for evidence about whether these animals possess markers of sentience such as sensory receptors for negative stimuli, brain regions that integrate information, sensitivity to local anesthetics or analgesics, the ability to make trade-offs between risks and rewards, and the ability to engage in self-protective behaviors.

The authors express very high confidence that lobsters have brain regions that integrate information, high confidence that they have sensory receptors for negative stimuli, and medium confidence that they have the capacity for associative learning. The authors also express very high confidence that astacid lobsters are sensitive to local anesthetics or analgesics but only low confidence that spiny lobsters have this sensitivity. This evidence might or might not establish that astacid and spiny lobsters are *likely* to be sentient. But it establishes that they have at least, say, a *one in ten* chance of being sentient.

Has our improved knowledge about lobsters resulted in improved treatment of lobsters? The evidence is mixed. On the one hand, we appear more willing to recognize the potential moral significance of lobsters; for instance, the UK government expanded the scope of its Animal Welfare (Sentience) Bill in 2021 to include invertebrates like crabs, lobsters, and octopuses. On the other hand, we have not yet followed this change in our beliefs and values with a comparable change in our behavior; for example, the supply of crustaceans for food has remained roughly stable in the UK for the past decade.

In the real world, nobody currently lives with a Neanderthal or a robotic human, since Neanderthals exist only in the past and robotic humans exist only in the future, if at all. But we do share the world with many beings who resemble us in some respects and not in others, including lobsters, insects, microbes, and plants. And in the future, the number and variety of such beings might only increase. In this chapter, we examined tools for estimating whether these beings matter and how much they matter in the face of substantial uncertainty. We can now examine what happens when we apply these tools in practice.

CHAPTER 4

Many beings might matter

IN JANUARY OF 2023, A startup called Innovafeed broke ground on a production site that can house more farmed animals than any other location in history. But the animals in question are not cows, pigs, or chickens. Instead, they are black soldier fly larvae. When the facility is fully operational, Innovafeed estimates that it will have the capacity to produce sixty thousand metric tons of fly larvae protein each year. And this facility is only one of many that will soon exist. According to one estimate, the insect farming industry may be worth nearly $10 billion by 2030, producing more than three billion metric tons of protein per year.

While information about the lives of farmed insects is limited, the evidence currently available raises troubling questions about insect welfare. At present, there are no laws governing the treatment of farmed insects. As a result, humans farm insects at very high densities. And while some insects prefer group living, others are at risk of stress, injury, disease, conflict, and cannibalism. We then slaughter insects via methods such as freezing, baking, roasting, microwaving, boiling, blanching, asphyxiating, and shredding. Some insects also become feed for farmed or companion animals, who sometimes eat them alive.

These questions are also pressing in cases where information about the lives of farmed insects is unavailable. Industrial animal agriculture has not gone well for the tens of billions of land animals and hundreds of billions of aquatic animals killed for food every year. This industry breeds, raises, and kills animals in a way that systematically prioritizes profit over welfare. And of course, farmed cows, pigs, chickens, and fishes are all vertebrates about whose moral significance we can be relatively confident. We have little reason to expect that farmers will exhibit more concern for black soldier fly larvae and other invertebrates.

While insect farming is the newest way in which humans kill insects in large numbers, it is far from the only way. We kill insects for silk, shellac, and carmine dye. We apply insecticides in our homes, schools, and offices. We kill quadrillions of insects every year with agricultural pesticides. And of course, human-caused environmental changes are affecting insect populations as well, causing some to contract and others to expand. Whether these outcomes are good or bad for insects depends at least partly on their quality of life, about which we know very little. Either way, the fates of countless insects now lie in our hands.

As a reminder, I am assuming here that consciousness, sentience, and agency are very likely jointly sufficient for moral standing. We can be confident that all beings who can consciously experience happiness and suffering and set and pursue their own goals have lives that matter to them. To be clear, these capacities might not all be *necessary* for moral standing; for instance, it might be that beings who can *unconsciously* detect positive or negative stimuli or set and pursue their own goals have lives that matter to them as well. But we can set that possibility aside for now and ask which beings might have these capacities.

I am also assuming here that we should extend at least some moral consideration to all beings with at least a *non-negligible chance* of having moral standing. Specifically, when determining the scope of the moral circle, the question that we should ask is not whether particular beings *do* matter, but rather whether they have at least a *one in ten thousand chance* of mattering, given the evidence available. If a robot or insect (or robotic insect!) has at least a one in ten thousand chance of having a life that matters to them, then they merit at least *some* consideration, even if only very little, in decisions that affect them.

The most important and difficult question is which beings might have *consciousness*. We know that many beings can detect helpful and harmful stimuli and set and pursue their own goals. What seems less clear is whether these beings have the capacity for subjective awareness in general, and whether it feels like anything for them to be detecting these stimuli and pursuing these goals in particular. If so, then these beings have consciousness, sentience, *and* agency. If not, then they lack consciousness—and for philosophers who hold that sentience and agency require consciousness, they lack sentience and agency as well.

My aim in this chapter is to survey beings who have at least a minimal capacity to detect helpful and harmful stimuli and set and pursue their own goals, and to ask whether they might have consciousness too. Of course, we should be cautious about taking this approach, since even if a being has all three of these capacities, we need to examine how these capacities interact before we can determine that this being can experience happiness, suffering, and other such welfare states. Still, when a being has all three of these capacities, we should at least take them seriously as a potential moral patient as we seek further evidence.

So, which beings might have consciousness? We can start by establishing a baseline. According to the New York Declaration on Animal Consciousness (organized by me, Kristin Andrews, and Jonathan Birch and signed by hundreds of experts in science, philosophy, and policy), the scientific evidence now supports a very strong chance of consciousness in all mammals and birds, as well as "at least a realistic possibility of consciousness" in "all vertebrates (including all reptiles, amphibians, and fishes) and many invertebrates (including, at minimum, cephalopod mollusks, decapod crustaceans, and insects)."

In keeping with arguments that Birch and I make elsewhere (including in this book), the declaration also calls for extending consideration to these animals:

> When there is a realistic possibility of conscious experience in an animal, it is irresponsible to ignore that possibility in decisions affecting that animal. We should consider welfare risks and use the evidence to inform our responses to these risks.

The aim of this statement is not to support any specific policy, but is rather to support the general idea of taking welfare risks for these animals seriously.

However, even if we take there to be a non-negligible chance of consciousness in all vertebrates and many invertebrates, we still need to examine the possibility of consciousness in other beings too. The category of carbon-based beings is large and diverse. It includes many other categories of invertebrates, such as annelids (earthworms, leeches, and more), nematodes (roundworms), and poriferans (sponges). It also includes many living beings who are not animals, such as plants (trees, grasses, and more), fungi (mushrooms, molds, and more), and single-celled microorganisms like bacteria and archaea.

The category of silicon-based beings has the potential to be even larger and more diverse. We already create AI systems for a variety of purposes, ranging from work to love to war. Many of these AI systems exist in physical space as robots, vehicles, or other embodied agents. Many others exist in virtual space as chatbots, avatars, or other "disembodied" agents. And in the future, a much wider range of silicon-based beings could exist in both physical space *and* virtual space, with some resembling animals, plants, fungi, and other carbon-based beings, and others taking brand-new forms that we can barely even imagine at present.

The future might contain *linked* minds, too. We might one day be able to link our brains together so that they can share mental states directly. This prospect is not as far-fetched as it may seem; for instance, an octopus can be described as having nine linked brains—one central brain and another, smaller, connected brain in each arm. Is it possible that a conscious subject resides in each brain *and* in the octopus as a whole? Now suppose that researchers one day link two octopuses together (not that they should!). Is it possible that a conscious subject now resides in each brain, in each octopus as a whole, *and* in the two octopuses as a pair?

The future might contain *mixed* minds as well. We can already create minds that are partly human and partly nonhuman, say by implanting nonhuman cells into humans or by creating animals with both human and nonhuman cells. We can also create minds that are partly carbon-based and partly silicon-based, say by implanting silicon chips into humans or animals or by creating beings with both carbon-based and silicon-based materials. And in the future, we might even be able to create minds that are partly physical and partly virtual, say by link-ing our brains with servers and storing memories or other mental states in the cloud.

How can we tell which of these beings, if any, are conscious? The most common answer involves a kind of *argument from analogy*. Each of us knows that it feels like something to be *us*. So, insofar as other beings are similar to us—behaviorally, anatomically, and evolutionarily—we can have confidence that it might feel like something to be *them*, as well. Of course, this method can be unreliable; for instance, since chatbots are designed to mimic our behavior, we should take behavioral similarities between humans and chatbots with a heavy pinch of salt in this context. But when we apply this method with care, it can be worthwhile.

To be clear, not everyone accepts this method—or any method—for determining which beings are conscious. Some experts doubt that we can have knowledge of other minds at all, since the only mind that each of us can directly access is our own. Suppose that you want to know how many voters support a particular policy, yet you have the ability to poll only a single voter. It would be hard for you to make a reliable estimate based on a sample size of one! Yet when you want to know how many beings are conscious and you have the ability to poll only a single mind—your own—how is this any different?

Without exploring this skeptical view in detail, I will make three observations about it. First, *if* we accept this skeptical view, then we should apply it to *all* other beings: other humans, other mammals, other vertebrates, and so on. Second, the upshot of this skeptical view is not that we should *deny* that it feels like anything to be other beings, but rather that we should be *uncertain* about this. And third, we should not simply exclude other beings from the moral circle in cases of uncertainty about whether it feels like anything to be them. Instead, we have a responsibility to proceed with caution and humility.

Skepticism aside, suppose that we at least endorse the main empirical claims in the New York Declaration on Animal Consciousness: The scientific evidence supports a very strong chance of consciousness in all mammals and birds as well as a realistic possibility of consciousness in all vertebrates (including reptiles, amphibians, and fishes) and many invertebrates (including cephalopod mollusks, decapod crustaceans, and insects). The question is then: What should we think about the vast multiplicity of other beings with whom we share the world now, and with whom we might share the world in the future?

Answering this question partly requires assessing how likely particular theories of consciousness are to be correct. However, this task is difficult due to ongoing disagreement and uncertainty about the nature of consciousness. At one end of the spectrum, many experts accept *inclusive* theories of consciousness. For example, some hold that consciousness is a basic property of all matter. Others hold that consciousness requires only, say, the capacity to process information, represent objects, or convert perceptual inputs to behavioral outputs via states like beliefs and desires. These theories imply that a wide range of systems—complex or simple, centralized or decentralized, and carbon-based or silicon-based—can be conscious.

At the other end of the spectrum, many other experts accept *exclusionary* theories of consciousness. For instance, some hold that consciousness requires particular kinds of higher-order thoughts (that is, thoughts about other thoughts), which, in turn, require particular kinds of cognitive unity and complexity. Others hold that consciousness requires particular kinds of chemical and electric signals, which, in turn, might require a carbon-based substrate at present. These theories imply that only a narrow range of systems—namely, particular kinds of complex, centralized, and/or carbon-based systems—can be conscious at present.

Many other experts strike a balance, holding that consciousness requires more than, say, simple information processing but less than, say, particular kinds of higher-order thoughts or chemical and electric signals. The exact requirements vary from view to view, but common requirements include capacities for perception, attention, learning, memory, self-awareness, and decision-making. And these capacities, in turn, plausibly require a moderate amount of cognitive complexity, cognitive centralization, and/or substrate-specificity. These theories thus imply that a moderately wide range of systems can be conscious at present.

How much confidence should we have in particular theories of consciousness? While this question is difficult to answer, I think we can say at least this much at this stage: Given ongoing disagreement and uncertainty about the nature of consciousness, we should approach this topic with caution and humility. And in my view, approaching this topic with caution and humility requires accepting that middle-ground theories of consciousness have at least a non-negligible chance of being correct and that other, more inclusive and exclusionary theories have at least a non-zero chance of being correct, given the evidence available.

Of course, making progress on this issue also requires estimating how likely particular nonhumans are to be conscious according to particular theories of consciousness. Yet this task is difficult as well, due to the limited and mixed evidence available. For example, we still have very little evidence about most insect species. And the evidence that we do have supports attributions of consciousness in some respects and not in others; for instance, honeybees can communicate the location of food through symbolic dance, yet they do not appear to groom their injuries, which is a common behavior in animals who can feel pain.

Our understanding of organisms like annelids, nematodes, sponges, plants, fungi, and microbes is even more limited and mixed. For example, nematodes have only about three hundred neurons each, yet they can still perceive and react appropriately to stimuli. Similarly, plants have no neurons at all, yet they can still perceive, learn, remember, communicate, and react appropriately to stimuli. We might not notice these kinds of organisms performing these kinds of behaviors, since, for instance, nematodes tend to be small and transparent, and plants tend to move relatively slowly. But these capabilities exist all the same.

Our understanding of AI systems is no better. AI systems are often black boxes: We know *that* they work without always knowing *how* they work. For example, humans created large language models to perform a single task: predict text. But we have limited knowledge about how they perform this task, and it might be that they perform it partly by developing new capabilities. Consider that evolution created us to perform a single task, too: reproduce. We then developed a wide range of capabilities, including consciousness, in the pursuit of this goal. What kinds of capabilities might AI systems now be developing?

Fortunately, researchers are starting to make progress on this question. For instance, in 2023 a team led by Patrick Butlin and Robert Long surveyed several leading theories of consciousness for indicators of AI consciousness related to perception, attention, learning, memory, and other such capacities. They then examined several leading AI systems to determine whether these indicators are present. They concluded that "no current AI systems are conscious" according to these theories. However, they also concluded that "there are no obvious technical barriers to building AI systems" that possess these indicators in the future.

HOW CAN WE PUT these reflections together to estimate how likely particular beings are to be conscious? We can start by noting that all of the beings discussed in this chapter have at least a *non-zero* chance of being conscious. There is at least a non-zero chance that an inclusive theory of consciousness is correct—that consciousness resides in, if not all matter, then at least all simple information-processing systems, representational systems, or systems that can convert perceptual inputs to behavioral outputs. And there is also at least a non-zero chance that all of the beings discussed in this chapter are conscious according to these theories.

For people who accept the no threshold view—which, as a reminder, holds that all non-zero risks merit at least some consideration—this conclusion is already noteworthy. All of the beings discussed in this chapter can—at least in a minimal sense—detect helpful and harmful stimuli and set and pursue their own goals. So if they have a non-zero chance of consciousness, then it might not take much extra work to show that they have a non-zero chance of sentience and agency, too. According to the no threshold view, this is all it takes to establish that they merit at least *some* consideration, even if only very little.

However, do these beings have a *non-negligible*—that is, at least a *one in ten thousand*—chance of being conscious? This question is harder to answer, because it requires us to make at least rough estimates about probabilities. Specifically, we would need to estimate how likely each theory of consciousness is to be correct, then estimate how likely each being is to be conscious according to that theory, and then put it all together to estimate how likely each being is to be conscious. Of course, this is difficult to do with any precision or reliability, but it can still be a useful way to test our assumptions and intuitions about other minds.

For people who accept the threshold view—which, as a reminder, holds that only *non-negligible* risks merit consideration—this kind of math might be necessary for estimating the scope of the moral circle.

Yes, an inclusive theory of consciousness *could* be correct. But if the probability that such a theory is correct is extremely low—say, only one in a quintillion—then showing that such a theory *could* be correct is not enough to show that this possibility merits consideration. Since I want my conclusions in this book to be broadly acceptable, I will focus on asking which beings pass this "non-negligibility test" for consciousness.

I personally feel ambivalent about this issue. On the one hand, it seems obvious to me that at least *some* of the beings discussed in this chapter fail the non-negligibility test for consciousness. I feel skeptical that the simple ability to process information, represent objects, or convert perceptual inputs to behavioral outputs is enough. I also feel skeptical that flowers, mushrooms, robots, or chatbots currently possess features required by more demanding theories. If asked to place a bet, I would bet a lot of money on the idea that consciousness requires a set of structures and functions that, at present, only carbon-based animals possess.

On the other hand, I think that it would be hubris for me to deny that there is at least a *one in ten thousand* chance that an inclusive theory of consciousness is correct. The fact that it feels like anything at all to be anyone at all is *astonishing*. If I were a robot, and you told me that slabs of meat experience happiness when their meat parts send chemical signals back and forth in the right kind of way, I would be skeptical about that, too. Yet I know from personal experience that at least one slab of meat *does* experience happiness. That makes me want to be at least somewhat skeptical about my own skepticism in this context.

On balance, I lean toward accepting that all of the beings discussed in this chapter do, in fact, pass the non-negligibility test for consciousness, since I lean toward accepting that inclusive theories of consciousness do, in fact, have at least a one in ten thousand chance of being correct. Many other philosophers lean this way, too; for example, on a 2020 survey of philosophers, 13.63% of respondents indicate that they accept or lean toward the idea that consciousness is a basic property of all matter. Given how difficult this topic is, I think that I should take this possibility seriously, my own skeptical reaction notwithstanding.

I also lean toward accepting that *many* of these beings pass the non-negligibility test for consciousness *whether or not* an inclusive theory of consciousness is correct. Even if simple informational systems, representational systems, and perceptual-behavioral systems lack consciousness, more complex systems with advanced and integrated versions of these capacities still have a good chance of being conscious. And many of the beings discussed in this chapter have a good chance of meeting this higher standard given the evidence available, including all vertebrates, many invertebrates, and many future AI systems.

In the category of carbon-based beings, we can debate whether, say, plants, fungi, and microbes meet this moderate standard. But it seems difficult to deny that many animals meet it. For example, even if we restrict ourselves to the taxa listed in the New York Declaration on Animal Consciousness (which I doubt we should do), there are likely to be quadrillions, if not quintillions, of animals alive at any given time in these taxa. Yes, many of these animals lack a cerebral cortex, higher-order thought, and other such features. But it would be premature to hold that these features are *extremely likely* to be necessary for consciousness.

Similarly, in the category of silicon-based beings, we can debate whether, say, current robots and chatbots meet this moderate standard. But it seems difficult to deny that, say, future robots and chatbots will. It is only a matter of time before we create silicon-based systems with advanced and integrated capacities for perception, attention, learning, memory, self-awareness, social awareness, language, reason, and more. And while these systems might lack our carbon-based substrate and evolutionary origin, it would once again be premature to hold that these features are *extremely likely* to be necessary for consciousness.

These reflections are unsettling for me. While it would take much more work to estimate exactly which beings pass the non-negligibility test for consciousness with any precision or reliability, I feel compelled to assume as a working hypothesis that all of the beings discussed in this chapter pass the test, and that many of them, including insects and future AI systems, *clear it by leaps and bounds*. However, since this assumption might be controversial, I will not insist on it here. Instead, I will simply insist that many of these beings, including insects and future AI systems, *pass the test*. That much seems impossible to deny in good faith.

At this point, we should also recall that some beings might merit consideration *whether or not* they pass the non-negligibility test for consciousness with a "one in ten thousand" threshold for non-negligibility. First, insofar as a more inclusive theory of moral standing might be correct, some beings might merit consideration even if they have only, say, *nonconscious* experiences or motivations. And second, insofar as a lower risk threshold might be correct, some beings might merit consideration even if they have only, say, a *one in ten million* chance of being conscious, sentient, agential, or otherwise morally significant.

I personally feel ambivalent about inclusive theories of moral standing in the same way that I do about inclusive theories of consciousness. On the one hand, it seems clear to me that only conscious beings matter for their own sakes. If plants, microbes, robots, or chatbots can detect stimuli or pursue goals but have no subjective awareness of anything at all, then I feel skeptical that it matters to them what happens to them.

On the other hand, it seemed clear to many past philosophers that only *humans or other rational beings* matter for their own sakes, and this view is now widely recognized as unacceptably exclusionary.

On balance, I lean toward accepting that inclusive theories of moral standing do, in fact, have a non-negligible chance of being correct. No matter how confident I feel about the basis of moral standing, I need to stay mindful of my place in the cosmic order. Had I been born in a different time or place, I would have had access to different information and arguments. And had I been born into a different species or with a different substrate, I might have had access to different forms of intrinsic value. I should take there to be a decent chance that my scientific *and* moral views are mistaken in light of these contingencies.

When I consider the possibility that extremely low risks merit consideration, I feel split as well. On the one hand, it seems plausible to me that I can permissibly neglect such risks in many cases. I should not devote my life to taking care of all the microbes who live on, and in, my body. On the other hand, it also seems plausible to me that I should consider such risks in at least *some* cases, particularly when the stakes are high. Again, even if Oppenheimer took there to be only, say, a one in ten million chance that testing the atomic bomb would destroy the planet, this risk still should have factored into his decision *somehow*.

On balance, I lean toward accepting that extremely unlikely risks do, in fact, merit consideration. I might not be required to drop everything to consider every single risk in every single situation in practice. When the stakes are low and my time, information, and rationality are limited, I might need to consider some risks while neglecting others, even if I recognize that they all merit consideration. Still, extremely unlikely risks *do* merit consideration in theory. And when the stakes are high and I have enough time, information, and rationality to assess extremely unlikely risks, I should do my best to consider them in practice too.

When I wrote in the introduction of this book that I feel troubled by my own arguments, this is part of what I meant. I set out to show that we should include insects and future AI systems in the moral circle, and yet I now feel compelled to include microbes, current AI systems, and many other beings, too. In fact, when I combine my view that these beings have a one in ten thousand chance of being conscious, my view that more inclusive theories of moral standing have a one in ten thousand chance of being correct, and my view that lower risks merit consideration too, this conclusion becomes *overdetermined* for me.

With that said, I will not assume that we should include all of these beings in the moral circle in this book, though I personally think that we should and will sometimes explore the implications of this idea. Instead, I will simply assume that we should include all vertebrates, many invertebrates (including insects), and many silicon-based beings (including advanced future AI systems), since I think that all reasonable moral and scientific views converge on this conclusion. I will then ask what we owe these beings in a world reshaped by human activity, and I will argue that the answer to this question is: *Uh oh*, we owe them *a lot*.

However, I will also argue that the details depend on many factors, including all of the questions that we asked in the previous two chapters. For instance, how much moral weight do insects and AI systems carry? Should we assign equal weight to all potential moral patients, or should we assign more weight to potential moral patients who are more likely to matter and who are likely to matter more? Moreover, should we accept that a large population of small individuals can carry more weight than a small population of large individuals overall, or should we deny that the moral weight of individuals can be combined in this way?

Similarly, what do moral agents owe moral patients in general? Is morality primarily about promoting welfare, respecting rights, cultivating virtuous characters, or cultivating caring relationships? Either way, how do relational and practical factors shape and limit our duties to distant others? Do we have a duty to prioritize distant nonhuman populations on the grounds that they carry more weight than human populations (assuming they do), or do we instead have a right to prioritize nearby human populations on the grounds that we have special relationships with each other and a greater ability to treat each other well (assuming we do)?

As I will argue, we can take solace in the fact that these moral, relational, and practical factors will shape our duties substantially. In her famous article "A Defense of Abortion," Judith Jarvis Thomson argued that abortion can be permissible in many cases even if fetuses have a right to life. We will similarly find that, say, killing insects and AI systems can be permissible in many cases even if they have a right to life. Yes, we might have a duty not to kill these beings *unnecessarily*. But we might still have a right to kill them in self-defense, in other-defense, as an unavoidable side effect of important activities, and for other reasons.

However, we can take only so much solace in such observations. As we will see, once the moral circle is expansive enough to include insects and future AI systems (to say nothing of microbes or current AI systems), human exceptionalism will be difficult to maintain. One way or another, nonhumans will matter a lot, *either* because large groups of small nonhumans matter a lot *or* because small groups of large nonhumans do. And while we might still be able to prioritize ourselves to an extent on moral, relational, or practical grounds, we might not be able to prioritize ourselves to nearly the extent that we do.

THE SCALE OF INSECT farming is difficult to grasp. In order to produce three billion metric tons of insect protein per year, the insect farming industry would need to kill roughly forty-five to fifty trillion insects or larvae per year. That would be nearly four hundred times more farmed insects per year than the total number of humans who have ever lived. It would also be more farmed insects per year than the total number of mammals killed by humans for food in the entire history of human civilization. If you lined up that many insects end to end, the line would stretch from Earth to the sun and back *more than three thousand times.*

Proponents of insect farming tout it as a sustainable alternative to traditional animal farming. In a limited sense, that might be true. In addition to harming and killing more than one hundred billion farmed vertebrates per year, animal farms are leading consumers of land, water, and energy, and they are also leading producers of waste, pollution, and greenhouse gas emissions. Indeed, animal farming produces an estimated 14.5% of all human-caused greenhouse gas emissions, partly because it releases high quantities of greenhouse gases and partly because it contributes substantially to deforestation, which does the same.

Even if we set aside concerns about insect welfare, insect farming is not the environmental savior that many believe it to be. At present, insect farming and traditional animal farming are complementary, not competitive, industries; for example, many insect farmers plan to use waste from traditional animal farms for insect feed, and many aquatic animal farmers plan to use insect protein for fish feed. The industry is also lobbying to allow land animal farmers to use insects as feed. By reducing the cost of animal feed, insect farming might produce an *expansion* of traditional animal farming.

Moreover, I have argued that we should *not* set aside concerns about insect welfare. There is a decent chance that the rise of insect farming will exploit and exterminate *hundreds of trillions of moral patients* in the coming decades, all for the sake of unclear benefits for humans. This risk merits serious consideration now, before this industry scales up any further. Yes, in a world with a vast number and wide range of potential moral patients, all food systems are risky. But some are riskier than others. And if insect farming is both massively and unnecessarily risky, then we might need to kill this industry in its larval stage.

CHAPTER 5

If we might be affecting you, we should assume we are

• • • • • • •

On September 26, 1983, Lieutenant Colonel Stanislav Petrov of the Soviet Air Defense Forces was on call at the Oko nuclear early-warning system when the system detected that multiple U.S. intercontinental ballistic missiles were headed toward the Soviet Union. The protocol was for Petrov to report this finding to his superiors, and the likely result of this report would have been a full-scale nuclear war, with an estimated 136–288 million humans dying directly and another 2 billion dying indirectly. But Petrov and his staff decided not to report the strike, since the signal might be a false alarm. They were right.

You might hope that if the fate of the world rested on your shoulders, you would rise to the occasion and save the world, too. But it can be hard to summon this kind of resolve. We all get sucked into our own little corners of the world; we have bills to pay, families to support, properties to maintain, careers to pursue, or other projects that consume our attention. So, when we find ourselves in situations that require us to see beyond our personal cares and concerns—to consider not only the local, individual stakes that feel concrete to us but also the global, collective stakes that feel abstract to us—it can be disorienting.

Nevertheless, if you knew that your next decision would determine whether, say, global nuclear war occurs, you would have a responsibility to rise to the occasion. Yes, deviating from protocol might be bad for you, but causing a nuclear strike would be bad for *everyone*, and you need to consider both of these risks when deciding what to do. And in fact, the same would be true if you merely *suspected* that your next decision would have this kind of large-scale impact. As we have seen, if an action has a non-negligible chance of harming moral patients against their will, you have a responsibility to consider this risk.

The question that we now face is: What would morality be like if this kind of situation were the norm? That is, what would morality be like if our *ordinary* decisions had the potential to affect billions, trillions, quadrillions, quintillions, or even sextillions of morally significant beings? In that scenario, we would need to engage in large-scale moral thinking as the rule, not the exception. And since this kind of moral thinking is difficult, we would need to build new structures in our lives—new kinds of roles, rules, habits, projects, relationships, and even societies—to assist us in thinking, feeling, and acting this way.

OUR OLD FRIEND DAVID, the factory manager, is still looking for a way to dispose of his waste. He now accepts that, when treating the waste is an option, it would be wrong for him to simply dump his untreated waste in a lake, thereby poisoning the lake and harming all the animals who live there. But since treating his waste involves a minor cost, David still prefers to find another solution if he can. His factory generates a hundred gallons of waste each week, and he wants to find a way to dispose of it all without poisoning lakes and harming animals and without paying for a chemical and physical treatment process.

Then, one day, David comes up with a possible loophole: Instead of dumping *a hundred* gallons of waste in *one* lake, he can dump *one* gallon of waste in each of *a hundred* lakes. David reasons that while a hundred gallons of waste is enough to poison a lake, one gallon of waste is not. So if he distributes his waste across a hundred lakes, then the waste will switch from being harmful to being harmless. Yes, this method might take extra time. But the lakes are all nearby and David enjoys spending time outdoors, so he still sees this method as less costly than treating the waste. He thus starts distributing his waste across the lakes.

However, David soon learns that many other factory managers are distributing their waste across these lakes as well. That means that David is not, as he previously thought, causing each lake to have *one* gallon of waste. Instead, David is causing each lake to have one *extra* gallon of waste. And while one gallon might not be enough to poison a lake, there is *some* number of gallons between one and a hundred that *is* enough. So each time David dumps his waste, there is a chance that his action is causing the waste in that lake to switch from harmless to harmful. Yet he decides to maintain his new routine anyway.

This example once again concerns the ethics of risk. In Chapter 3 we discussed moral and scientific uncertainty about *moral standing*—about which features are necessary for a being to be a moral patient, and about which beings have these features. We now need to consider moral and scientific uncertainty about *right action*—about which features are necessary for an action to be right, and about which actions have these features. As we will see, only when we examine uncertainty about moral standing and right action together can we start to appreciate how difficult morality might really be.

One day, as David is making his rounds dumping the waste, he notices several rabbits standing at the edge of a lake, looking out into the water. He walks around the lake to take a photo, and when he gets closer, he sees what they see: a rabbit, stuck in the shallow part of the lake. She can keep her head above water for now, but it will only be a matter of time before she drowns. David realizes that he has only two options in this situation. Either he can wade into the shallow part of the lake, saving the rabbit but delaying his rounds and destroying his clothes. Or he can continue with his plan for the day, allowing the rabbit to die.

Should David save the rabbit? He feels skeptical. First of all, he feels skeptical that he has a general duty to help others where possible. Yes, many smart people hold that we should help others where possible. But many smart people also reject this idea; instead, they hold that we should avoid harming others unnecessarily, that we should help those whom we have harmed where possible, and that we can pick and choose when we help others beyond that. David sides with these latter philosophers; he thinks that he has a right to do as he pleases, as long as he minimizes the unnecessary harm that he causes along the way.

David also feels skeptical that he has a specific duty to help this rabbit. Yes, *if* his actions caused the rabbit to be in this situation, *then* he would have a duty to intervene. But David feels skeptical that his actions had this effect. The rabbit probably would have gotten stuck whether or not the factory managers poured their waste into the lake. And in the unlikely event that she *did* get stuck because of the waste, that probably would have happened whether or not David participated. Yes, the rabbit has a life that matters to her, and to others. Her death will be a tragedy. But David feels confident that his hands are clean.

However, even if David has reason to be skeptical about having a duty to help this rabbit, he also has reason to be skeptical about his skepticism. After all, he might be wrong about ethics. Yes, he feels confident that non-interference matters more than assistance. Intuitively, *causing* someone to suffer seems worse than *allowing* them to. But when David considers how complex these issues are, how many smart people disagree with him, and how many sources of bias he has (it would be convenient for him if morality were less demanding!), he should take there to be at least a decent chance that his confidence is misplaced.

David might also be wrong about his impacts. Yes, he feels confident that his past actions made no difference for this rabbit, and that his present action will have no other significant impacts. But there is at least a decent chance that his past actions *did* make a difference—that his gallon of waste triggered a tipping point and the accumulated waste then trapped the rabbit. There is also a decent chance that his present action will shape his character and relationships moving forward—for instance, that helping the rabbit would reinforce the virtue of kindness and that neglecting her would reinforce the vice of callousness.

As we have now discussed at length, these possibilities can be morally significant. Recall that the risk principle holds that non-negligible risks merit consideration, and that a one in ten thousand chance of a morally significant impact counts as non-negligible. This principle implies that if there is at least a one in ten thousand chance that David is wrong about ethics—that he does, in fact, have a general duty to help others where possible—then he should factor that possibility into his decision-making somehow. Specifically, he should feel at least *some* pressure to save the rabbit *whether or not* his actions caused her predicament.

Now recall that the harm reduction principle holds that we should reduce and repair the unnecessary harm that we cause, and that we should cultivate habits, relationships, and other structures that support this work. This principle, together with the risk principle, implies that if there is at least a one in ten thousand chance that David is wrong about his impacts—that his past actions did, in fact, cause this predicament, or that his current action will, in fact, shape his character and relationships in morally significant ways—then he should factor these possibilities into his decision-making in the same kind of way.

Of course, even if David feels pressure to save the rabbit for either or both of these reasons, other factors might be relevant to his decision, too. For example, is there a risk that saving this rabbit would somehow be bad for her, her family, or other animals? Is there a risk that it would somehow be bad for David, his family, or other humans, beyond the minor cost of delaying his rounds and destroying his outfit? And if saving the rabbit *does* involve morally significant trade-offs, which factors take priority? Still, whether or not David should feel *decisive* pressure to save the rabbit, he should feel at least *some* pressure to save her.

As this discussion illustrates, there are important parallels between uncertainty about moral standing and uncertainty about right action. In both cases, we face uncertainty both about values (that is, which features are necessary for moral standing and right action) and about facts (that is, which beings and actions have these features). In both cases, this uncertainty is likely to last for a long time, due to the complexity of these issues and the limitations of our human perspectives. And in both cases, acting responsibly in the face of this uncertainty requires factoring morally significant possibilities into our decision-making.

There are also important interactions between these forms of uncertainty. Suppose that there is a decent chance that sextillions of current and future insects and AI systems can suffer *and* that our actions and policies are affecting them. In this case, my arguments imply that we should feel at least some pressure to not only extend them moral consideration but also help them where possible and avoid harming them unnecessarily. Needless to say, that would constitute a massive expansion of our moral responsibilities. In effect, we would be assuming that we have at least a weak duty to *actively assist sextillions of beings* where possible.

How should we feel about the idea that we might have a duty to assist such a large number of beings? A positive spin is that it represents philosophy at its best, since it reflects our ability to use evidence, reason, and empathy to improve our beliefs, values, and practices over time. We start with a plausible and widely accepted set of premises; in this case, these premises involve our duty to reduce harm and consider risk. We then reach a provocative and revisionary set of conclusions; in this case, these conclusions involve at least a weak duty to help *and* avoid harming insects *and* AI systems in other nations *and* future generations.

In contrast, a negative spin—which is a heightened version of the objection that we considered in Chapter 3—is that this conclusion represents philosophy at its *worst*, since it reflects our tendency to allow simple, abstract principles and arguments to lead us to disaster. Yes, it might seem plausible that we should reduce harm and consider risk. But if we simply assume that everything that *might* matter *does* matter and that everything that *might* happen *will* happen, then we might end up assuming that so much matters and that so much will happen that we neglect what really *does* matter, and what really *will* happen.

As in Chapter 3, this objection is well-taken. Specifically, if we assume that we have even a weak duty of assistance to sextillions of current and future nonhumans, then we might decide that we should systematically prioritize these beings over ourselves. For example, we might decide that we should invest in infrastructure changes that *could* benefit future insects and AI systems instead of infrastructure changes that *would* benefit humans here and now. And of course, it would be bad if we systematically sacrificed *actual* benefits for *actual* moral patients for the sake of merely *possible* benefits for merely *possible* moral patients.

However, we can make the same kinds of replies here as in Chapter 3. First, the risk of *underestimating* the significance of distant possibilities can sometimes be worse—more likely *and* more severe—than the risk of *overestimating* their significance. Consider that we spent decades neglecting climate change because many humans viewed it as purely imaginary. We now recognize that this negligence was a regrettable, if convenient, mistake. We might now feel tempted to neglect our impacts on many nonhumans for similar reasons. But we should be open to the possibility that this negligence would be a mistake as well.

Second, there is no need for us to take a single-issue, "either-or" approach to global ethics. When climate ethicists argue that we should improve our impacts on future humans, they are not arguing that we should systematically prioritize future humans over current ones. Instead, they are arguing that we should seek co-beneficial policies for everyone where possible and that we should prioritize thoughtfully where necessary—where this plausibly means making at least minor sacrifices for the future while still meeting our needs at present. Why not assume that a similar balance can be struck with humans and nonhumans?

In previous chapters, I noted that we should keep an open mind about these issues in part because morality is a marathon, not a sprint, and in part because the risk of human bias is pervasive. These points are especially relevant here. First, before recoiling at the idea that we might have duties of assistance to sextillions of nonhumans, we should keep in mind that this idea has no direct or obvious implications about how we should live. That will depend on a wide range of interconnected moral, relational, and practical factors, and there is no need to assume the "worst"—say, the inevitability of excessive self-sacrifice—at the outset.

Second, even if we have a strong intuition that we have no duties of assistance to sextillions of nonhumans, we should keep in mind how many biases might be implicated in this intuition. We have trouble empathizing with beings whom we have an incentive to harm or neglect. We have trouble empathizing with beings who are neither like us nor near us. We have trouble imagining low-probability and high-magnitude impacts. And we have a tendency to view morality through a single-issue, "either-or" lens rather than through a multi-issue, "both-and" lens. These biases, and many others, place us at an enormous deficit here.

Fortunately, we have tools that we can use to overcome this deficit, which we will explore over the next two and a half chapters. For now, the key idea is that we should not simply dismiss distant, large-scale possibilities for nonhumans as purely imaginary. If there is a decent chance that these beings matter and that our actions and policies are affecting them, then we should feel at least some pressure to help them *and* avoid harming them where possible. The question will be: How can we seek co-beneficial policies for all actual and potential stakeholders where possible, and how can we prioritize thoughtfully where necessary?

DAVID STILL FEELS UNEASY about poisoning the lakes, so he makes two final changes to his waste-disposal strategy. First, instead of distributing his waste across *one hundred* lakes, David decides to distribute it across *one thousand* lakes. That means that instead of taking a one in a *hundred* chance of poisoning each of a *hundred* lakes, David will take a one in a *thousand* chance of poisoning each of a *thousand* lakes. While the expected harm might be the same in both cases (David can expect to poison one lake per round of distribution either way), he still likes the idea of imposing ten times less risk on any particular lake.

Second, instead of dumping his waste in lakes that contain fishes, annelids, arthropods, and other "macroscopic" beings, David can dump his waste in lakes that contain nematodes, rotifers, tardigrades, and other "microscopic" beings. This change has a similar effect, since, David estimates, the populations in the new lakes are ten times larger than the populations in the old lakes but are also ten times less likely to matter. Specifically, he estimates that the animals in the old lakes have a one in a *hundred* chance of mattering and that the animals in the new lakes have only a one in a *thousand* chance of mattering.

This case once again raises questions about the moral significance of extremely unlikely possibilities. Yes, the one thousand lakes strategy has a one in a thousand chance of impacting these microscopic organisms, which is a non-negligible possibility. And yes, these microscopic organisms have a one in a thousand chance of mattering, which is a non-negligible possibility. But it follows that the one thousand lakes strategy has only a *one in a million* chance of impacting beings who matter, which, David thinks, is a negligible possibility. This approach thus reduces the risk of harm to what David takes to be a tolerable level.

This case also raises questions about individual and collective responsibility. In general, when an individual participates in a harmful activity, we can ask about responsibility both at the individual level and at the collective level. In this case, for instance, we can ask whether David should risk poisoning the lakes as an individual, and we can also ask whether the factory managers should be poisoning the lakes as a collective and whether David should be participating in this activity as an individual. How can we assess the one thousand lakes strategy at each level, and what follows for the permissibility of this strategy?

At the individual level, a lot depends on whether a one in a million risk counts as negligible. If so, then the one thousand lakes strategy might, indeed, be better than the one hundred lakes strategy. Both strategies provide David with a minor benefit: They allow him to avoid the minor cost associated with treating his waste. However, the one hundred lakes strategy also creates a non-negligible risk of large-scale harm, which makes it wrong according to the harm reduction and risk principles. In contrast, the one thousand lakes strategy creates only a negligible risk of large-scale harm, which makes it all upside.

However, if a one in a million risk counts as non-negligible, then the one thousand lakes strategy might *not* be better than the one hundred lakes strategy. For people who address risk by estimating impacts, these strategies are equivalent, since whether you take a one in a *hundred* chance of poisoning each of a *hundred* lakes or a one in a *thousand* chance of poisoning each of a *thousand*, the expected result is the same: one poisoned lake. And for people who address risk by avoiding worst-case scenarios, the one thousand lakes strategy is even *worse* than the one hundred lakes scenario, since it risks poisoning *ten times as many lakes*.

How can we assess these strategies at the collective level? Many philosophers believe that we can be collectively responsible for harms that we collectively produce and that we can be individually complicit in harms that we participate in producing. Suppose that each of a hundred assassins gives one-tenth of a lethal dose of poison to each of a hundred victims. The victims all die, yet none of the assassins seems to be individually responsible for any of the deaths. According to this view, however, the assassins can be collectively responsible for the deaths *whether or not* any particular assassin is individually responsible too.

How does this view about collective responsibility apply to the one thousand lakes strategy? When David and the other factory managers pursue this strategy, they collectively poison one thousand lakes. And since the beings in these lakes have a one in a thousand chance of mattering, poisoning these lakes means imposing a one in a thousand risk of harm on these beings. Thus, the factory managers are collectively responsible for producing a non-negligible risk of large-scale harm, and David is individually complicit in this risky collective activity *whether or not* his individual actions produce a non-negligible risk as well.

Many consequentialists accept a variation of this idea as well. Specifically, according to *rule consequentialism*, we should perform actions that would have good consequences if everyone performed them, and we should avoid performing actions that would have bad consequences if everyone performed them. And in this case, whether or not it would be bad for *any particular* factory manager to dump their waste, it would definitely be bad for *every* factory manager to do so, since they would collectively poison all the lakes and kill all the animals. Thus, rule consequentialism implies that David should treat his waste.

Many rights theorists accept a variation of this idea too. According to the *universal law principle*, we should perform actions that would be acceptable to us if everyone performed them, and we should avoid performing actions that would be unacceptable to us if everyone performed them. (This view is similar to rule consequentialism, except the focus is less on the value of the outcome and more on the moral and rational integrity of the agent.) And in this case, David would definitely find it unacceptable if *every* factory manager dumped their waste. Thus, the universal law principle implies that David should treat his waste too.

Many virtue and care ethicists are similar. They hold that our actions can express morally significant character traits and place us in morally significant relationships whether or not they make a difference in particular cases. They also hold that when we knowingly and willingly become "part of the problem," we express and promote a vicious, uncaring kind of self-regard, whereas when we knowingly and willingly become "part of the solution," we express and promote a virtuous, caring kind of other-regard. Thus, these theories imply that David should treat his waste as well—or, at least, that a virtuous and caring agent would do so.

Finally, as we have seen, ethicists of all kinds also focus on the *shared structures* within which we think and act together. In general, when we face a problem, it helps to ask not only how we can solve this problem at present but also how we can avoid this problem in the future. In this case, for example, whether or not David and the other factory managers should treat their waste at present, they should still develop shared structures—such as an infrastructure for treating their waste, subsidies for treating their waste, or penalties for dumping their waste—that make it easier for everyone to treat their waste in the future.

We can thus see that many paths lead to the same destination: The one thousand lakes strategy is, in one way or another, problematic. Yes, *if* morality is entirely about the impacts of individual actions, and *if* a one in a million risk is negligible, *then* there is nothing problematic about this strategy. But every other theory that we have surveyed implies that there *is* something problematic about it. The details might differ, but the general idea that we should participate in helpful rather than harmful patterns of activity, and that we should cultivate habits, relationships, and other structures that support this work, is robust across many theories.

This creates a strong case for skepticism about the one thousand lakes strategy. As we have seen, when we assess our actions at the level of theory, we should give at least some weight to each theory that has a decent chance of being correct. And in this case, even if David is confident that morality is entirely about the impacts of individual actions *and* that a one in a million risk is negligible, it would still be premature for him to be *certain* about this view at this stage. He should give at least *some* weight to at least *some* of these other theories, and he should thus feel at least *some* pressure to treat his waste in the spirit of caution and humility.

Additionally, when we assess our actions at the level of practice, we should note that different moral theories are mutually supporting. And in this case, even if David remains confident that his preferred kind of consequentialism is correct in theory, he might not always be able to assess the impacts of his actions reliably in practice. In these moments, he might find it useful to ask what would happen if *everyone* performed this action in this situation, as a kind of heuristic. While this heuristic is by no means perfect, it can still be a good way to select better, less risky actions in cases where impact assessments are hard to conduct.

Moreover, David might not always be able to deliberate about what to do, and even when he does deliberate, his character traits, relationships, and other structures are likely to have a pervasive influence on his actions too. Thus, David might find it useful to cultivate character traits, relationships, and other structures that positively reinforce the kinds of actions that he generally hopes to perform, as a kind of supplement to his methods of deliberation. Once again, this strategy is by no means perfect, but it can still lead to better actions in cases where deliberation is either impossible or, at least, insufficient.

If enough factory managers accepted enough of these ideas—either because they viewed them as reasonably likely to be correct in theory or because they viewed them as useful tools for selecting good actions in practice—then this collective action problem would be solved. A critical mass of factory managers would commit to treating their waste, they would develop a shared norm of treating their waste, and they would develop shared structures for treating their waste. That would be enough to create a tipping point for the entire community, and as a result, the factory managers would all be fine, and the lakes would all be clean.

This leaves David, and all of us, in an uneasy place. As we will see, we all regularly perform actions that have an extremely small chance of producing an extremely large impact, in part via our contribution to collective actions. When we consume more than our fair share of resources or produce more than our fair share of pollution, our actions might be extremely likely to benefit us and extremely unlikely to harm the world. But if they *did* harm the world, they would harm it *a lot*. And if *everyone* performed these actions, we would *definitely* produce this harm. That raises the question whether many ordinary actions are wrong.

Furthermore, even if we reject the idea that we should treat others well *collectively*, we should still accept that we should treat others well *individually*, and this is already enough to warrant radical moral circle expansion in some cases. Suppose that you work at NASA, and you need to decide whether to fund a mission that would send insects to live on Mars. In this case, even if all of your other decisions in life have only a negligible chance of helping or harming distant future insects, this decision might still have a non-negligible chance of doing so. If so, then these insects clearly merit consideration in this context.

And of course, opportunities to help or harm other, closer nonhumans will be more common. Suppose that you make food-procurement decisions for the New York City government. In this case, even if your *personal* procurement (what you purchase for yourself) has only a negligible chance of helping or harming the many nonhuman victims of our food system, your *professional* procurement (what you purchase for millions of public meals per day) might still have a non-negligible chance of doing so. If so, then the many nonhuman victims of our food system clearly merit consideration in this context.

And if, as I have argued in this chapter, we should aspire to treat others well individually *and* collectively, then this kind of moral circle expansion should be the rule, not the exception. Yes, when you decide what to eat for lunch today, the probability that your personal lunch order will help or harm the many nonhuman victims of our food system might be extremely low. But it might still be non-zero. And either way, the probability that *our* orders will help or harm these beings is much higher. For these and other reasons, the many nonhuman victims of our food system merit consideration in this ordinary context too.

This conclusion might seem overwhelming. But I think that we should accept it anyway. It follows naturally from the idea that we should reduce harm and consider risk both individually and collectively. And as we will now discuss in more detail, if we pursue co-beneficial solutions where possible, prioritize thoughtfully where necessary, and build shared structures that support this work, then we can achieve and sustain higher levels of care for everyone *without* excessive self-sacrifice. I think that we should accept that we share this responsibility rather than reject the plausible and widely accepted ideas that lead to this conclusion.

On February 24, 2022, Russia invaded Ukraine. This invasion caused tens of thousands of civilian casualties and hundreds of thousands of military casualties, making it the deadliest European invasion since World War II. Volodymyr Zelenskyy, the president of Ukraine, fought back with support from the United States and NATO, and

Vladimir Putin, the president of Russia, responded by threatening a nuclear attack on Ukraine. Fortunately, the international community managed to avoid nuclear brinkmanship in the months following these threats. But the threat of a nuclear exchange remains unacceptably high.

This series of events was a wake-up call for everyone who assumed that the days of nuclear threats were behind us. Nearly eighty years have passed since the United States detonated two nuclear weapons in Japan, but we should not assume that this hiatus will continue indefinitely. The world has witnessed not only several near misses but also steady proliferation during this period. At present, nine countries possess more than twelve thousand nuclear warheads, and other countries might soon join the nuclear club as well. Of the many humans who maintain these systems, how many are like Stanislav Petrov?

Meanwhile, we continue to face other global threats too, including the threat of pandemics, climate change, and totalitarianism. And new technologies not only introduce new threats but also amplify old ones. For example, AI could increase the risk of pandemics by facilitating the creation of harmful pathogens. It could increase the risk of climate change by contributing to industrialization. It could increase the risk of totalitarianism by contributing to surveillance and military technology. And it could increase the risk of nuclear war by automating processes that previously required input from humans like Petrov.

We might take comfort in the idea that we are unlikely to face the kind of decision that Stanislav Petrov faced. Unfortunately, many of us are not so lucky. Millions of humans work in governments and industries that directly interact with these technologies. And while the rest of us might not hold such positions, we can still make a difference for these global threats—or, at least, we can still participate in patterns of behavior that do so—through ordinary actions such as eating lunch or using the internet. As we will see, whether we—and others—survive this new age will depend in part on how we comport with this new reality.

CHAPTER 6

We might be affecting many beings

IN JUNE 2022, THOUSANDS OF cattle died due to heat and humidity in Kansas. These deaths led to widespread coverage about the impacts on the beef and dairy industry. But mostly lost in this coverage were reflections on the impacts for *the cattle*, each of whom died in a way that we would never wish for ourselves or our loved ones. Also lost in the coverage were reflections on the contribution that the beef and dairy industry makes to methane emissions, deforestation, and other drivers of the climate crisis that will increase the frequency and intensity of this kind of extreme weather event in the future.

These cattle are not the first farmed animals to die in this kind of extreme weather. In 2021, for example, hundreds of thousands of farmed animals died in a heat wave in the Pacific Northwest, and later that year, hundreds of thousands more died in floods in the same region. Climate change is, moreover, not the only global threat that humans and farmed animals currently face. Millions of farmed animals have died due to bird flu, COVID-19, and other diseases, either because the animals became sick or because farmers killed them to suppress the spread of disease or destroy excess inventory.

Moreover, farmed animals are not the only animals who are caught up in this kind of cycle of harm. Deforestation, development, and the wildlife trade kill trillions of wild animals per year, and they also contribute to pandemics, climate change, and other global threats that imperil us all. In 2019 and 2020, the Australia bushfires killed an estimated three billion wild animals, and the Amazon wildfires killed an unknown number of wild animals. And when wild animals survive such disasters, they become more vulnerable to ordinary threats like hunger, thirst, illness, injury, and human violence and neglect.

In the future, a similar dynamic might extend to other nonhumans as well. The development, deployment, and scaling of AI has the potential to harm an increasingly large and diverse nonhuman population, and it also has the potential to contribute to biosecurity threats, nuclear threats, and other global threats that imperil us all. Fortunately, we still have a chance to learn from our mistakes with animals and take a better path with both animals and AI systems. But learning this lesson requires accepting that human and nonhuman fates are now linked, due in part to the pervasive impacts of human activity on the rest of the planet.

Our moral faculties are outdated. We naturally perceive the direct effects of small-scale interactions as morally significant much more easily than the indirect effects of large-scale interactions. When ten people kill a hundred people here and now, we naturally track the entire causal sequence, and we naturally experience it as morally significant. When ten million people in one place and time contribute to environmental changes that kill a hundred million people in another place and time, the level of harm might be much higher, yet our ability to track this causal sequence and experience it as morally significant is much lower.

However, we now live in *the Anthropocene*, a geological epoch in which human activity is a dominant influence on the planet. While we still have morally significant effects via direct and small-scale interactions, we increasingly have morally significant effects via indirect and large-scale interactions, too. Every day, billions of humans consume land, water, energy, and other inputs at an alarming rate, and we produce pollution, carbon dioxide, methane, nitrous oxide, and other outputs at an alarming rate as well. These and other large-scale patterns of behavior will affect countless lives moving forward.

Thus, our new predicament is that, in addition to having a *high* degree of confidence that our actions make a *small* difference *directly*, we should now have a *low but non-zero* degree of confidence that our actions make a *large* difference *indirectly*, via their contribution to large-scale patterns of behavior. We thus need to balance a concern for the kinds of small, direct, likely impacts that we naturally perceive as significant with a concern for the kinds of large, indirect, unlikely impacts that we fail to naturally perceive at all. The possibility of ethical action in the modern world depends on our ability to strike this balance.

This predicament is at the heart of *the ethics of the Anthropocene*. In the last chapter, I argued that we have a responsibility to increase the probability that our actions will help others and decrease the probability that our actions will harm others. I also argued that we should accept a pluralist moral framework for doing this work, by aspiring to perform helpful individual actions, participate in helpful collective actions, and cultivate character traits, relationships, and other structures that support this work. We will now see that in the Anthropocene, we owe these responsibilities to a vast number and wide range of beings.

The Anthropocene increases the number and variety of beings to whom we can have duties in two ways. First, humans are *impacting* more beings than ever before, via new technologies and globalized social, political, and economic systems. Second, humans are *creating* more beings than ever before, via new technologies and large-scale uses of carbon-based and silicon-based beings. And when we combine these ideas, we can see that humans are also creating beings *via* our impacts, since human-caused global changes will shape what kinds of beings can exist and what kinds of lives they can have.

Consider each of these points in turn. First, humans are impacting more beings than ever before. Agriculture, deforestation, development, and other activities are radically reshaping the planet. They have this effect directly, by converting natural spaces into artificial spaces and affecting many individuals in the process. They also have this effect indirectly, by consuming natural inputs faster than the planet can create them, producing natural outputs faster than the planet can absorb them, and altering species and ecosystems faster than anyone can adapt, in ways that have the potential to affect nearly everyone.

These impacts connect us across space and time, often in ways that involve deep inequities. High-income humans in industrialized nations in the second half of the twentieth century and the first half of the twenty-first century are disproportionately responsible for climate change, yet low-income humans in non-industrialized nations and in future generations may be disproportionately impacted. Clearly, these inequities are morally significant. When one group of humans is harming another in foreseeable and avoidable ways, we have a responsibility to reduce and repair this harm where possible.

These impacts also connect us across biology and materiality, often, again, in ways that may involve deep inequities. Humans are disproportionately responsible for global threats like climate change, yet nonhumans will be disproportionately impacted. And since the world contains more invertebrates than vertebrates, we can likewise say that vertebrates are disproportionately responsible for global threats like climate change, and that invertebrates will be disproportionately impacted. These interspecies inequities, while less commonly discussed than intraspecies ones, are morally significant too.

Humans are also *creating* many more beings than ever before. We already create new beings through breeding and genetic modification. We breed farmed animals to produce as much meat, milk, and eggs as possible. We breed lab animals to develop diseases that we see as useful for research, such as cancer. We breed companion animals to have features that we see as cute, such as compressed noses. Should we create these animals in these ways at all? And if we do create them, do we have special duties to them given our responsibility for their existence and circumstances, and given their distinctive needs and vulnerabilities?

Humans are also creating increasingly sophisticated silicon-based beings. As we have seen, we already share the world with robots that provide us with work or care, with programs that create essays or images for us, and with avatars that talk, play, or fight with us in video games. And in many cases, we treat these beings in ways that would be clearly unacceptable if they were moral patients. Yes, we might be confident that, say, chatbots lack consciousness, sentience, agency, and other such features now. But can we be *certain* about that? And even if so, can we *remain* certain about that as these technologies develop?

Humans are also creating *chimeras*. As we have seen, researchers are creating animals with both human and nonhuman genetic material for research and transplantation. In the future, we might also create chimeras with both carbon-based and silicon-based "genetic" material for cognitive or physical enhancement. As before, we can ask: Should we combine beings in these ways at all? And if we do combine them, how should we treat these chimeras? For example, if we create human-pig or human-robot chimeras, then to what extent should we treat them like humans and to what extent should we treat them like pigs or robots?

Finally, humans are also creating many beings *via* our impacts. Human-caused climate change will impose new evolutionary pressures on animals, causing some species to contract, other species to expand, and other species to develop new traits. In this respect, the Anthropocene will blur the line between domesticated and wild animals, as well as the line between captive and free-living animals, since human activity will now be partly shaping the evolution of many species (making many species at least partly domesticated) *and* limiting the freedom of many individuals (making many animals at least partly captive).

Similar considerations apply to AI systems as well. We currently create all AI systems directly. That might not always be true, since AI systems might one day become capable of reproduction, and if and when they do, humanity will no longer be *directly* responsible for the existence of all such beings. But even at this point, humanity will still be *indirectly* responsible for the existence of all such beings, since we will have created the original AI systems, will have empowered them to reproduce, and will have shaped the environment in which they reproduce. In these respects, no AI system can ever be fully free of human influence.

The upshot is that human activity is now potentially affecting a *vast number* and *wide range* of potentially morally significant beings across species, substrates, nations, and generations. We already impact quadrillions of animals each year, and the more we expand the number and variety of potential moral patients and our power to potentially impact them, the more we will expand the scale of potentially morally significant impacts of our activity. Simply put, *sextillions* of nonhumans might soon merit consideration, including insects, AI systems, and other nonhumans in other nations and future generations.

Of course, there is a limit to how many nonhumans we can consider in our daily lives. Picture an ordinary college student named Sinan. He studies philosophy, works as a research assistant, runs a student organization, and spends time with his friends, family, partner, and roommates. Everything is going well, but Sinan is stressed. Not only does he need to work full-time to keep up with all his commitments, but he also needs to make major decisions about his future. For example, should he go to law school, medical school, or a PhD program? And should he aspire to build a family with his partner, or should he take a different path?

Sinan aspires to treat everyone who might matter with respect and compassion—his undergraduate thesis is about intergenerational ethics, and his student organization advocates for nonhuman rights—but with everything happening in his life right now, his capacity for moral deliberation is in short supply. Like all of us, Sinan performs thousands of actions each day, and he has sextillions of potential stakeholders for each action. Even under the best of circumstances, it would be impossible for him to assess how all of his actions might affect all of these beings. And these are definitely not the best of circumstances.

However, Sinan can—and, I believe, should—still accept that his potential impacts on these beings merit consideration. In general, when we say that a possibility merits consideration, the idea is not that we should always, or even often, perform the mental action of considering it. Instead, the idea is that we should factor it into our decision-making *somehow, where possible.* And in this case, even if Sinan lacks the ability to consider his potential impacts on these beings every time he performs an action, he still has the ability to factor these potential impacts into his decision-making in several ways that, at this point, should be familiar.

First, of course, Sinan can consider his potential impacts on these beings *sometimes.* He routinely makes small-scale decisions about how to treat insects, AI systems, and other such beings. When he interacts with ants or chatbots in his apartment, he can decide whether to treat them with kindness or cruelty. Moreover, each career path available to Sinan has the potential to lead to a different role in society. For all he knows now, he might one day be routinely making large-scale decisions about how to treat insects, AI systems, and other such beings as well, and his decisions now might influence his decisions later on.

Second, Sinan can also consider these beings when selecting the rules that he follows in everyday life. Suppose that he sees an ant in his apartment once every few months. In this case, he has no need to assess how to treat each individual ant. Instead, he can assess this issue once and then treat his decision—relocate the ant—as a general policy. Additionally, suppose that he has no idea whether his individual behavior makes a difference for large, distant nonhuman populations at all. In this case, he can still aspire to be part of the solution—say, by supporting companies that treat these populations well where possible—as a general policy.

Third, Sinan can consider these beings when selecting the habits, relationships, and other structures in his life. In the same way that individual decisions can eventually become general policies, general policies can eventually become personal habits: At some point, there will be no need for Sinan to think about how to treat individual ants or about how to make consumer decisions at all, because the relevant behaviors will simply flow out of him in the relevant situations. And if Sinan builds community with people who maintain similar policies and habits, then his community will reinforce these behaviors as well.

If Sinan expands his moral circle in these ways, will he always treat nonhumans well? Almost certainly not: His ability to consider his impacts is still limited, and his decisions, policies, habits, and relationships will likely lead him astray in many cases. But he can still treat nonhumans *better*. He can improve the probability that his actions will help and avoid harming nonhumans. He can treat nonhumans with respect and compassion *whether or not* his actions make a difference for them. And he can set an example that influences others—including his own future self—to expand their moral circles in similar ways.

As we will see in the next chapter, the same is true for humanity as a whole. At present, there is a limit to how many nonhumans our species can consider. Even if, say, distant future microbes matter and our actions are affecting them, we might lack the knowledge, power, and political will needed to assess—much less improve—our impacts on them. But like Sinan, we still face an open future. And if our species expands our moral circle now, then we might not only treat many beings better at present but also build the knowledge, power, and political will that our successors will need to treat many other beings better in the future.

How can we pursue policies that help as many individuals as possible now while developing the ability to help a larger number of individuals in the future? I think that the One Health policy framework is a useful model for this kind of policy-making. The UN Food and Agriculture Organization describes One Health as an "integrated approach" that recognizes links among "the health of animals, people, plants and the environment," with implications for "food security, sustainable agriculture, food safety, antimicrobial resistance (AMR), nutrition, animal and plant health, fisheries, and livelihoods."

The core idea of One Health—that by recognizing the links among human, animal, and environmental health, we can improve all three at the same time—is strong because it provides us with a practical strategy for helping many nonhumans at present while developing the ability to help many other nonhumans in the future. After all, even if we disagree about what to do in cases where human and nonhuman interests conflict, we can still work together in cases where our interests align. We can also work together to build shared institutions and infrastructures that expand opportunities for alignment in the future.

However, while the core idea of One Health is strong, our current applications of this framework have several limitations that blunt its impact. First, we tend to focus on the links between human and nonhuman *health* while neglecting related links between human and nonhuman *welfare*, *rights*, and *justice*. Second, we tend to focus on the value that nonhuman health has for *us* while neglecting the value that it has for *nonhumans*. Third, we tend to focus on the health of *"natural"* nonhumans like animals, plants, and ecosystems while neglecting the health of *"artificial"* nonhumans like chatbots, robots, and "infosystems."

If we preserved the strengths of the One Health framework while removing the limitations, how might our applications of this framework be different? What kinds of links would we be able to reveal, and what kinds of policies would we be able to create? I think that we would see that human activity is impacting a vast number and wide range of carbon-based *and* silicon-based nonhumans, thereby contributing to global threats that imperil health, welfare, rights, *and* justice for us *and* them. I also think that in many—though, admittedly, not all—cases, revealing these links can empower us to respond to these threats.

Consider first why animals matter from an expanded One Health perspective. Humans are intentionally killing more than one hundred billion captive animals and hundreds of billions of wild animals per year for food, research, and other purposes. This treatment is clearly bad for these animals. Factory farmed animals, who make up at least 90% of all farmed animals, have particularly bad lives. We often breed them to grow as big as possible as fast as possible, mutilate them without anesthesia, keep them in intensive confinement, and kill them on industrial disassembly lines that prioritize efficiency over welfare.

Our treatment of these animals also contributes to global threats like disease outbreaks. Factory farms typically maintain large populations of animals in toxic environments and routinely administer antibiotics and antimicrobials to stimulate growth and suppress disease. This practice creates ideal conditions for novel pathogens to develop and spread. Animal agriculture also contributes to deforestation. As a result, it decreases forested biodiversity and increases interactions between humans and wild animals, which contributes to the spread of zoonotic diseases both within and beyond wild animal populations.

Our treatment of these animals likewise contributes to global threats like climate change. Animal agriculture, particularly beef and dairy agriculture, not only consumes high quantities of land and water and produces high quantities of waste and pollution but also emits high quantities of carbon dioxide, methane, and nitrous oxide. It also, again, contributes to deforestation, which releases carbon dioxide into the atmosphere and diminishes the planet's ability to capture and store carbon dioxide in the future. As a result, animal agriculture is responsible for an estimated 14.5% of global human-caused greenhouse gas emissions.

These global threats then imperil humans and nonhumans alike by exposing us to diseases, fires, floods, and other disasters, as well as by amplifying ordinary threats like hunger, thirst, illness, and injury. Nonhumans can be particularly vulnerable during these disruptions. During a pandemic, we might kill them because we see them as "pests," because we see them as sources of a treatment, or because we lack the ability to care for them. And during an extreme weather event, we might kill them because we see them as "invaders," because we see them as sources of food, or, again, because we lack the ability to care for them.

To break this cycle, humans need to include animals in health and environmental policy by reducing our use of them as part of our global-threat mitigation efforts and by increasing our support for them as part of our adaptation efforts. That means phasing down harmful industries like factory farming and phasing up plant-based alternatives. It also means phasing up research and advocacy about helping animals, education and employment opportunities for helping animals, and resources and infrastructure for helping animals, so that humans will be better able to care for other animals in the future.

The problem, of course, is that violence against animals is deeply entrenched in human society. Our social, legal, political, and economic structures treat animals as objects that are here for us rather than as subjects who are here for themselves. Additionally, human and nonhuman oppressions are linked, since, for instance, we rationalize oppression of humans partly by comparing them with nonhuman animals who are presumed to be "lesser than" because of perceived cognitive or physical differences. These dynamics, and the links between them, are part of why improving our treatment of animals is so important and difficult.

Now consider why AI systems matter from an expanded One Health perspective. If and when AI systems become moral patients (whether or not you perceive them as such, as you do with your roommate Dara), humans might eventually use them at even vaster scales than we do with nonhuman animals. For instance, we might build a vast number of virtual worlds, and we might populate each of these worlds with a vast number of virtual beings for research, education, or entertainment. If we do, and if the beings who populate these worlds matter, then these practices might constitute oppression at an unprecedented scale.

AI development already produces risks and harms as well. Humans are already witnessing the effects of algorithmic bias. When we train AI systems with human data, they learn from the best *and* worst of humanity, and they produce surprisingly thoughtful *and* shockingly biased outputs as a result. And when we train AI systems with a *curated* set of words and images, they still learn from the best and worst of the *curators*. For instance, if the curators accidentally center white people in their inputs, then AI systems will likely do the same in their outputs. Either way, we risk amplifying human biases.

As AI development advances, questions about AI safety and alignment will be increasingly pressing. In general, the more powerful AI systems become, the more impactful their beliefs, values, and other dispositions will be. If sufficiently powerful AI systems have anything less than *ideal* beliefs and values, the result could be disaster. And of course, ensuring that AI systems have ideal beliefs and values is a difficult task. It requires not only knowing which beliefs and values are ideal but also instilling them in AI systems perfectly. It might also require getting it right the first time, since we might not have a second chance.

These threats can imperil humans and nonhumans alike. One possibility is that humans retain control of AI systems and use them to advance our own bad ends, for instance by establishing a permanent totalitarian state that locks us into an oppressive status quo. Another possibility is that humans lose control of AI systems and they become either antagonistic or indifferent toward us. A classic, if simplistic, example involves a company that deploys a powerful AI system to maximize paperclip production, and then the AI system gradually converts all matter—including humans, animals, and plants—into paperclips.

To mitigate these risks, we need to include AI systems in our policies as well. In particular, we should reduce our use of AI systems as part of our global-threat mitigation efforts and increase our support for AI systems as part of our adaptation efforts. That means slowing down AI development so that we can take the time needed to ensure that future AI systems have good beliefs and values if they exist at all. It also means speeding up research and advocacy around human, animal, *and* AI welfare, rights, and justice so we can consider *all* the potential stakeholders of our practices—and so AI systems can do the same.

Will humanity be able to learn from our mistakes with other animals in time to avoid repeating this pattern with AI systems (in addition to breaking the pattern with other animals, which is already hard enough)? Possibly. On one hand, AI is still at an early stage of development and integration into society. And preventing a harmful system from taking hold is easier than dismantling it once it does. On the other hand, even at this early stage, humanity is already dependent on AI. And our incentives to speed up and reap the rewards of AI are in tension with our incentives to slow down and mitigate the risks.

But even if our chances of success are low, we should still do what we can. At present and in the near future, our uses of nonhumans are not only increasing risks and harms for them but also increasing the risk of global threats like pandemics, climate change, nuclear war, and permanent totalitarianism for everyone. And to the extent that such disasters occur, they will threaten humans and nonhumans alike, both directly and indirectly. They will also involve deep inequities, since a small percentage of stakeholders will be disproportionately responsible—and a large, non-overlapping percentage will be disproportionately impacted.

Of course, this is not to say that we should extend full rights to animals and AI systems overnight. That would be impossible, and even if it were possible, it might be undesirable. After all, it will take a lot of work to build the conditions necessary for extending full rights to animals and AI systems without subjecting humans to famine, poverty, life in the Matrix, death at the paperclip factory, and other such risks as a result. Instead, my claim is simply that by considering human and nonhuman welfare, rights, and justice holistically, we can identify co-beneficial solutions for everyone to the extent that such solutions exist.

Fortunately, there are many practical steps that we can take to make progress in this regard. For example, the global community can use financial and regulatory policies to reduce support for factory farming and increase support for alternatives, while ensuring a just transition for humans who currently depend on factory farming for food or income. The global community can also use these kinds of policies to slow down AI development so that we can take the time that we need to develop moral, legal, and political frameworks that can protect humans, animals, and AI systems alike from risks associated with this technology.

Meanwhile, the global community can use infrastructural policies to pursue co-beneficial adaptations for humans, nonhumans, and the environment. For instance, we can install bird-friendly glass on new buildings, build wildlife corridors on new transportation systems, and build urban green spaces with feeding stations, water stations, and managed nonhuman populations. And if and when AI systems are sufficiently likely to be moral patients, we can likewise build shared environments that allow humans, animals, and AI systems to live together in relative harmony, both within and across physical and virtual spaces.

What about individual actors like Sinan, the overworked college student? As we have now discussed at length, they can—and, I think, should—support these collective efforts where possible, both because they *might* be able to make a difference in some cases and because they can aspire to be part of the solution either way. By extending moral consideration to everyone who might matter, by developing habits that express respect and compassion for all, and by contributing to shared structures that do the same, individual actors can play their part in allowing everyone to live together in relative harmony as well.

However, the idea of living together in *relative* harmony is key. There will likely always be conflicts both within and across species, substrates, nations, and generations. Thus, we will likely never be able to build a world in which *all* potential moral patients can flourish. Still, insofar as co-beneficial solutions are available, examining our impacts on everyone holistically will improve our ability to identify them. And insofar as co-beneficial solutions are unavailable, examining our impacts on everyone holistically will improve our ability to set global priorities thoughtfully, as we will see in the next chapter.

GIVEN HOW MANY THREATS we face, it can be easy to motivate ourselves by imagining negative futures that we want to avoid. Unfortunately, many of these futures are easy to imagine because they flow naturally from the present. Factory farming is still on the rise, and if we continue on this path, then our food system will continue to kill animals, cause disease outbreaks, and cause extreme weather events—and then humans will continue to kill or abandon animals when these crises occur, either because we view these animals as "pests" or "invaders" or, as with the cattle in Kansas, because we lack the ability to properly care for them.

Many other negative futures are harder to imagine, but thinking about them can still be motivating. Science fiction is full of cautionary tales involving AI systems. In many such stories, humans create silicon-based beings so that we can exploit or exterminate them for our own selfish purposes, and we then lose control of these beings and *they* start exploiting or exterminating *us*. Of course, AI systems might or might not cause catastrophic harm in reality, and even if they do, the details are likely to be much different. Either way, the stakes of AI development for humans, animals, *and* AI systems are extremely high.

But as motivating as these negative futures can be, it can also help to imagine positive futures that we want to pursue. When I imagine such futures, I think about cities teeming with human and nonhuman life, supported by governments that consider everyone equitably and infrastructures that accommodate everyone equitably, including during (fortunately less frequent) crises. I also imagine selective, strategic deployments of AI systems that treat humans, animals, *and* AI systems with respect and compassion, because we developed AI systems slowly and cautiously and with everyone—including them—in mind.

Imagining the future in a constructive way requires striking a balance between optimism and pessimism so that we know what to pursue *and* what to avoid. It also requires striking a balance between radical and moderate visions so that we can think beyond our current trajectory without succumbing to naive utopianism *or* defeatism. In this chapter, I argued that we might be impacting many beings, and that we can improve these potential impacts in part by seeking co-beneficial solutions for humans, animals, and AI systems. In the next chapter, we can ask how to set priorities when co-beneficial solutions are unavailable.

CHAPTER 7

Against human exceptionalism

IN JANUARY 2022, DAVID BENNETT, a fifty-seven-year-old man in Baltimore, became the first human to receive a heart from a genetically modified pig. Bennett reportedly had heart failure and a "history of not complying with medical instructions," which rendered him ineligible for a human heart. However, he was eligible for an experimental surgery involving xenotransplantation, that is, cross-species transplantation. He selected this alternative, and so he received a heart from a pig with genetic modifications that medical researchers developed to increase the compatibility between porcine hearts and human bodies.

Bennett survived the procedure, and the initial results seemed promising. According to Dr. Bartley Griffith, a professor of surgery at the University of Maryland School of Medicine, "We were incredibly encouraged by his progress," given that "his heart was strong, almost too strong for his frail body, but he had a strong will to live." Despite these positive signs, Bennett died two months later. However, his medical team remained encouraged, since his autopsy indicated that he died due to heart failure, not organ rejection. As a result, his doctors reportedly considered his experimental surgery an early success.

This surgery was the first of its kind, but it will not be the last. As we saw in the last chapter, humans are increasingly breeding, raising, and killing nonhumans, including human-nonhuman chimeras like the animal used in this procedure, for food, research, transplantation, and other human purposes. And while these practices might sacrifice nonhumans, many humans think that the sacrifice is worth it provided that we can improve the technology so it fulfills its potential. As bioethicists Arthur Caplan and Brendan Parent put the point: "Animal welfare certainly counts, but human lives carry more ethical weight."

This assessment is too simple, however. We are now in a position to see that *even if* some lives carry more ethical weight than others, we should *still* reject human exceptionalism. Given the scale of nonhuman suffering in the world and the extent of human complicity in nonhuman suffering, we have strong duties to nonhumans. Yes, we should improve human lives and societies, in part because doing so is likely to benefit humans and nonhumans alike in the long run. But we should prioritize nonhumans more than we do. And as our ability to support nonhumans increases, our responsibility to support them will increase as well.

Human exceptionalism takes many forms, but for present purposes we can understand it as the ethical view that humans matter more than nonhumans, and that we owe humans more than we owe nonhumans, both individually and collectively. According to this view, nonhumans might matter a lot, and we might have strong duties to them. But we matter more than they do overall, and we have stronger duties to each other than we have to them overall. Thus, in a world with scarce resources and abundant need, we should prioritize humans over nonhumans in cases where we appear to have conflicting needs.

Some philosophers argue that humans merit priority because of our *intrinsic* value. For example, Shelly Kagan argues that we should prioritize humans because humans have stronger and more significant interests than nonhumans. Other philosophers argue that humans merit priority because of our *relational* value. For example, Baruch Brody argues that we should prioritize humans because we have stronger and more significant bonds within our species than beyond it. According to these views, if a house is burning down and I can save either you or a fruit fly, then I should default to saving you.

Importantly, not everyone accepts that we should prioritize some moral patients over others based on these intrinsic or relational differences. As we have seen, some philosophers believe that we should give equal weight to all moral patients no matter how strong their interests happen to be, and no matter how strong our bonds with them happen to be. According to these views, if a house is burning down and I can save either you or a fruit fly, then I should either flip a coin or search for another way to break the tie between you ("Well, the human is easier to catch, but the fruit fly is easier to carry . . .").

However, I want to grant for the sake of discussion that we should prioritize some moral patients over others based on these intrinsic and relational factors, and I want to argue that we should reject human exceptionalism anyway. Even if humans have stronger interests in some cases, nonhumans have stronger interests in other cases. And even if we have stronger bonds within our species in some cases, we have stronger bonds beyond our species in other cases. The upshot will be that we should still prioritize ourselves to an extent, but we should not prioritize ourselves to nearly the extent that we currently do.

First, why might we think that humans have stronger and more significant interests than nonhumans? Many philosophers accept this view in part because they think that humans have the capacity for more complex and varied motivations than nonhumans. As a human, you have the power to pursue a wide range of goals in life, since you can think about what to do and how to live. In contrast, a fruit fly has the power to pursue only a narrow range of goals in life, since they simply do what feels natural from moment to moment, without ever stopping to think about whether they should try to act or live differently.

Similarly, many philosophers think that humans have the capacity for more intense and prolonged experiences than nonhumans. According to this view, since you have a more complex brain than a fruit fly (humans have about four hundred thousand times as many neurons as fruit flies), you can experience more happiness, suffering, and other such states at any given time. And since you have a longer lifespan (the average human lifespan is about eighty years and the average fruit fly lifespan is about eighty days), you can also experience more happiness, suffering, and other such states over the course of your life.

Why might these differences matter? As we discussed in Chapter 1, many philosophers think that moral patients with higher capacities for welfare in these respects have *more at stake in life*. For instance, it would be very bad for me to torture you for the rest of your life, in part because I would be causing a very large amount of frustration and suffering and preventing a very large amount of satisfaction and happiness. In contrast, it would be less bad for me to torture a fruit fly for the rest of their life (while still, of course, being very bad if the fruit fly is a welfare subject), since they have less at stake in all these respects.

Similarly, it would be very bad for me to kill you early in your life, even if I do so painlessly. After all, even if killing you causes no frustration or suffering at present, it might still prevent a lot of satisfaction and happiness in the future. In contrast, it would once again be less bad for me to kill a fruit fly early in their life, since they have less at stake in these respects. Of course, this is not to say that I should never harm or kill humans or that I can always harm and kill fruit flies. Instead, it is only to say that the bar for killing humans is higher at least in part because we have higher capacities for welfare, and thus more at stake in life.

However, as we have seen, humans might not always have more at stake than nonhumans for these reasons. First of all, humans might not always have more complex or varied motivations than nonhumans. We all lack the capacity for rational deliberation early in life, some of us lose this capacity later in life, and some of us never develop it at all. Meanwhile, while nonhuman animals appear to lack this capacity, many nonhuman animals still have the capacity to set and pursue goals in a complex, intelligent manner. The idea that we always have a larger number or wider range of desires and preferences is thus likely false.

Moreover, in the future, many AI systems might have the capacity for more complex and varied motivations than human *and* nonhuman animals. Many experts believe that we will eventually witness an intelligence explosion, with new kinds or degrees of digital agency that exceed human agency as much as, if not more than, human agency exceeds other kinds of animal agency. At this point, humanity would no longer be the "apex agents." Instead, we would simply be one kind of agent among many, hoping that "higher" agents treat us with a kind of care that we rarely showed to "lower" agents during our time in power.

Likewise, humans might not always have more intense or prolonged experiences than nonhumans. Humans in persistent vegetative states might have a low capacity for happiness and suffering at any given time despite having complex brains, and humans near the end of their natural lives might have a low capacity for these states across time despite having long lifespans. Meanwhile, many nonhuman animals might have a very high capacity for happiness and suffering; for example, African elephants have three times as many neurons as humans, and immortal jellyfishes have indefinite lifespans.

And once again, in the future, many AI systems might have the capacity for more intense and prolonged experiences than human *and* nonhuman animals. In the same kind of way that we might witness an intelligence explosion, we might also witness a *consciousness* explosion, with new kinds or degrees of subjective experience that vastly exceed our own. At that point, some AI systems might be like imperfect gods from our perspectives: They might be, if not *all*-knowing, *all*-powerful, and *all*-experiencing, then at least capable of kinds of knowledge, power, and experience that we can admire but never truly comprehend.

Moreover, even if humans have more at stake than nonhumans at the *individual* level, nonhumans can still have more at stake than humans at the *population* level. There are many more animals than humans in the world, and in the future, there might be many more silicon-based beings than carbon-based beings in the world. If the moral weight of lives can be aggregated, then these differences matter. Specifically, even if an individual human carries more weight than an individual insect or chatbot, a sufficiently large number of insects or chatbots can still carry more weight than a sufficiently small number of humans.

And of course, even when humans have more at stake than nonhumans *in general*, the nonhumans might still have more at stake than the humans *in particular cases*. Suppose that I have to decide between sparing you from a paper cut that will hurt *a bit* for *a day* and sparing a pig from a laceration that will hurt *a lot* for *a year*. You might be at risk of suffering more in general, but the pig is still at risk of suffering more in this particular case. And ultimately (with apologies for the mild injury that you might now endure), what matters for my decision is which one of you has more at stake in this particular case.

This analysis bodes poorly for the idea that humans always, or even often, have the strongest and most significant interests. Humanity is one species among many, and nonhuman animals outnumber us by a ratio of at least one hundred million to one. Moreover, we routinely impose *major* burdens on *large* nonhuman populations in exchange for *minor* benefits for *small* human populations. The "harm" of eating plant-based meat instead of factory-farmed meat, for instance, pales in comparison with the harm—for more than one hundred billion vertebrates and one trillion invertebrates each year—of being factory farmed.

Of course, one might challenge this analysis by challenging some of the moral assumptions that it makes. For example, one might challenge the idea that we should give more weight to an advanced AI system than to a human by challenging the assumption that we should give more weight to more complex beings in general. One might also challenge the idea that we should give more weight to a sufficiently large population of insects or chatbots than to a sufficiently small population of humans by challenging the "rebugnant" assumption that we should give more weight to larger populations in general.

However, this objection is unpersuasive. First of all, it seems clear that we should give more weight to more complex beings and to larger populations in at least some contexts, and for at least some purposes. If sparing you from a laceration would prevent more suffering than sparing a fruit fly from a laceration, then this difference clearly matters. Similarly, if sparing a million humans from lacerations would prevent more suffering than sparing one human from a laceration, then this difference clearly matters as well. At the very least, the idea that these differences matter is plausible enough to merit consideration.

Moreover, challenges to human exceptionalism arise no matter what. For example, if we deny that an advanced AI system matters more than a human, then should we also deny that a human matters more than a fruit fly? And if we deny that a large population of insects or chatbots matters more than a small population of humans, then should we also deny that a large population of humans matters more than a small population of advanced AI systems? It is hard to imagine non-arbitrary principles that imply that we matter more than larger populations of smaller beings *and* more than smaller populations of larger ones.

One might also challenge this analysis by challenging the empirical assumptions that it makes. For example, one might insist that humans will always be more complex in relevant respects than nonhumans, since humans are currently more complex in relevant respects than both animals and AI systems, and since AI development will slow or stop before that changes. One might also insist that the human population will eventually be larger than the nonhuman population, since humans will eventually settle other planets and solar systems—and will leave potentially morally significant nonhumans behind when we do.

However, this objection is unpersuasive too. At this point in time, it seems plausible that some nonhumans will, in fact, be more complex than some humans, since companies are currently racing to develop artificial general intelligence, and since they seem at least as likely to succeed as to fail, if not more likely. It also seems plausible that the nonhuman population will, in fact, remain larger than the human population in the future, either because *nobody* will settle other solar systems or because *humans and nonhumans alike* will do so. At the very least, these empirical claims are plausible enough to merit consideration as well.

Additionally, whether these empirical claims are correct is not merely a matter for us to *predict*. It is also a matter for us to *decide*. Suppose that we do, in fact, produce a future in which we matter the most. Why will we have done so? Will we have produced this future because we aspired to build a better world for all, and because we discovered that expanding the human population while contracting the nonhuman one happened to be the best way to do that? Or will we have produced this future because we never aspired to build a better world for anyone other than ourselves at all? We will return to this issue below.

In the meantime, we can conclude that humans are unlikely to have stronger and more significant interests than nonhumans as a general matter. Yes, we might have more at stake than nonhumans do in many cases, particularly in cases involving local, small-scale conflicts. But we might also have less at stake than nonhumans do in many cases, particularly in cases involving global, large-scale conflicts. So, if our species is warranted in prioritizing ourselves over nonhumans as a general matter, then the reason must have less to do with our intrinsic value and more to do with other, more relational considerations.

Relational considerations are common in priority-setting decisions. For example, many philosophers take it for granted that we should prioritize ourselves and our families on relational grounds. I should take care of myself before I take care of you, and I should also take care of my family before I take care of yours. And if we should have this kind of partiality for families, perhaps we should also have this kind of partiality for species, substrates, nations, and generations, since we might have closer bonds within these groups than beyond them, and since equal weight for all members of all groups would be too demanding.

We can unpack this argument by noting that it combines several ideas that have less to do with how much everyone has at stake and more to do with how we relate to everyone. First, we might think that we should prioritize helping others when we have stronger bonds with them. I have a special duty to take care of my family by providing them with food, water, shelter, affection, and other such goods on a daily basis in part because I have special bonds of care and interdependence with them. In contrast, I have no such duty to the vast majority of other moral patients in part because I have no such bonds with them.

Second, we might think that we should prioritize helping others when we have a greater ability to help them. For example, given the small size of my family and my strong personal bonds with them, I have the knowledge, power, and motivation necessary to achieve and sustain high levels of support for them on a daily basis. In contrast, given the large size of the global population and my weak personal bonds with the vast majority of moral patients beyond my family, I clearly lack the knowledge, power, and motivation necessary to achieve or sustain similarly high levels of support for the rest of the world.

Third, we might think that we should prioritize helping others when doing so produces indirect benefits. For example, I need to take care of myself to an extent to be able to take care of others sustainably. Similarly, my family members need to take care of each other to an extent to be able to take care of others sustainably, individually as well as collectively. On this view, I can be warranted in prioritizing myself and my family to an extent not only because of the direct benefits for us, but also—assuming that we do, in fact, rise to the occasion and take care of others more as a result—because of the indirect benefits for others.

If we accept that we can prioritize our families to an extent for some or all of these reasons—which, again, not everyone does—then we might accept that we can prioritize our species to an extent for some or all of these reasons, too. As with our families, but at a larger scale and to a lesser extent, we have special bonds with fellow humans. We also have the knowledge, power, and motivation necessary to achieve and sustain higher levels of support for humans than for nonhumans. And insofar as we have a duty to support nonhumans, we need to take care of each other to be able to collectively take care of them sustainably.

As we discussed in Chapter 2, we might likewise think that we can prioritize our nation and generation to an extent for these reasons. Yes, distant moral patients might have much more at stake than nearby ones. But we still have stronger bonds with nearby moral patients, we still have a greater ability to treat them well, and we still *need* to treat them well to be able to collectively treat distant moral patients well, in the spirit of One Health. In these respects, we might think that we have much stronger relational duties to, say, nearby humans than to, say, distant nonhumans in spite of (and, partly, *because* of) our smaller population.

However, even if we accept that these relational differences exist and matter, we should still reject human exceptionalism in several important respects. First, none of these relational differences supports prioritizing my family *by any means necessary*. Yes, I might have a duty to feed my family before I feed your family. But should I *kill* your family so that I can feed human flesh to my family simply because they have a marginal preference for human flesh over, say, rice and beans? Of course not. The fact that I have a duty to prioritize my family to an extent does not mean that I have a right to instrumentalize yours.

Second, none of these factors supports prioritizing my family *at any cost to others*. Yes, I might have a duty to feed my family before I feed your family. But if I have the power to either save my family or save a hundred, thousand, or million other families during a global crisis, then should I still prioritize my family in this case? Plausibly, at some point the answer switches from yes to no. In general, even if my bonds with everyone matter, the stakes for everyone matter too. And when these relational and non-relational factors conflict, the moral significance of the stakes for everyone can take priority in at least some cases.

Third, I have special relationships beyond my family as well. Not only do I have personal relationships with many other humans, ranging from friends to colleagues to neighbors, but I also have impersonal relationships with *many, many* other humans via our shared structures. And as we have now discussed at length, these relationships can generate special duties too. For example, I can have a special duty to help strangers in many cases because of my complicity in social, political, and economic systems that contribute to pandemics, climate change, and other global threats that imperil them unnecessarily.

These considerations all extend to our relationships beyond our species, substrate, nation, and generation as well. For example, even if we have a duty to prioritize humans to an extent, that does not mean that we have a right to kill nonhumans simply because we prefer animal-based foods over plant-based foods. It does not mean that we have a right to neglect nonhumans even when we have the ability to help them sustainably. And it does not mean that we have a right to neglect our relational duties to nonhumans, duties which stem not only from our bonds with them but also from our impacts on them.

Indeed, when we consider the extent to which human activity impacts nonhumans in the Anthropocene, we might find that we have a duty to prioritize humans less and nonhumans more *both* because of the high stakes for nonhumans *and* because of our complicity in their predicament. As we have seen, we currently kill *trillions* of nonhumans per year unnecessarily, thereby contributing to global threats that imperil current *and* future humans *and* nonhumans. These practices place us in morally significant relationships with nonhumans all over the world, generating an extremely strong duty to help them where possible.

However, the fact remains that we can sustain higher levels of support for, say, nearby humans than for, say, distant nonhumans. The fact also remains that we need to take care of each other to be able to take care of others sustainably. Thus, what follows from my argument so far is not that we should prioritize members of other species, substrates, nations, and generations over ourselves. ("The U.S. should devote a majority of our current GDP to supporting simulated insects in other solar systems and future millennia!") Instead, what follows is that we should prioritize these populations as much as we can, while still taking care of ourselves.

Of course, our capabilities might change over time. Recall Sinan, the college student we met in the last chapter. Sinan is now in law school, and he recently got married and had a baby (yes, with his college partner!). He still wants to help others as much as possible, but he also needs to work eighty hours per week to earn his degree and support himself and his family. Given these realities, it seems plausible that Sinan has both a right and a duty to devote the vast majority of his resources to himself and his family at this stage in life, so he can take care of them all, and so he can support his, and their, education and development.

But now suppose that ten years later, Sinan is a partner at a top law firm. He earns $10 million per year, and he estimates that $1 million per year is more than enough to erase his debts, invest in savings, and establish a reasonably comfortable life for himself and his family. At this point, it seems plausible that Sinan both *can* and *should* help others much more than he was previously able to do. Specifically, it seems plausible that Sinan both can and should devote the vast majority of his resources to individuals outside of his family, while still devoting enough to himself and his family to ensure that they can live reasonably well.

Can we expect Sinan to make this altruistic choice when the time comes? The evidence is mixed. Humans have a greater capacity for altruism in times of abundance than in times of scarcity—though, of course, plenty of humans are still selfish in times of abundance. So a lot depends on how Sinan approaches his time as a student and associate. If he pursues success by any means necessary, then he might be more likely to remain selfish when he succeeds. But if he pursues success in a way that helps others as much as possible along the way, then he might be more likely to remain (at least partly) altruistic when he succeeds.

I think that we can tell the same kind of story about humanity as a whole. Like Sinan as a student, our species is still at an early stage in our education and development. We thus have both a duty and a right to prioritize ourselves at present, because we need to take care of ourselves and invest in our education and development. However, if and when we develop the ability to devote a majority of our resources to nonhumans sustainably, we might have a duty to do so at that point. And if we help nonhumans as much as possible along the way, then we might be more likely to make this altruistic choice when the time comes.

This last point is worth emphasizing. There is a path dependence to how history unfolds. Our successors will inherit the beliefs, values, practices, traditions, institutions, and infrastructures that we leave behind, and this inheritance will partly determine whether they have the knowledge, power, *and* motivation necessary for higher levels of altruism. By considering nonhumans as much as possible now, we not only help more individuals at present but also increase the probability that our successors will do the same in the future, at higher scales. By neglecting nonhumans unnecessarily now, we have the opposite effects.

Of course, this path dependence can take other forms too. Suppose that we eventually create AI systems whose knowledge and power vastly exceeds our own. In that scenario, we might find that our behavior influences our *nonhuman* successors too. Specifically, if we demonstrate through our behavior that moral agents should treat potential moral patients well *whether or not* they happen to be made out of the same materials, then AI systems might be more likely to inherit these new, more inclusive and egalitarian values. Otherwise they might be more likely to inherit our current exclusionary and hierarchical values.

I think that we can draw three general conclusions about our duties to nonhuman populations at present and in the future. First, we should prioritize nonhumans much more than we do at present, both for our sakes and for theirs. As we saw in the last chapter, when we apply an expanded One Health framework to our interactions with animals and AI systems, we can identify co-beneficial solutions for humans and nonhumans alike, including ending factory farming, deforestation, and the wildlife trade; and developing AI systems slowly and cautiously, with both AI safety and AI welfare in mind.

Second, if we do this work well, then our successors (either humans, nonhumans, or both) might eventually be able, willing, *and* morally required to prioritize other kinds of beings over themselves. The more we invest in our own knowledge, power, and altruism, the more we empower our successors to live in service of others. And if and when our successors become capable of living *primarily* in service of others—

devoting a majority of their resources to projects that support members of other species, substrates, nations, or generations—they will have a responsibility to make this altruistic choice at that point.

Third, however, taking this virtuous path requires telling ourselves new stories about the meaning, purpose, and value of human existence. From our current perspectives, a world in which humans are required to prioritize nonhumans might seem like a loss for our species. But if we really did create that world, then that would be a *win* for our species. Yes, we might have a duty to live primarily in service of others at that point, but the fact that we had this duty would be a vindication of all our hard work over the course of many generations, and of our resulting capacity to care for ourselves and others at the same time.

Ultimately, human exceptionalism is rooted in the idea that our species has a special kind of significance. I have argued that this idea is partly right and partly wrong. We *do* have a special kind of significance, because for better or worse, our actions will determine what the future holds for humans and nonhumans alike. However, this does not mean that we have a right to prioritize ourselves over nonhumans no matter what. Instead, it means that we have a duty to work toward a future in which those in power extend all due consideration to everyone who might matter—even if that means prioritizing others over themselves.

IN CONVERSATIONS ABOUT HUMAN exceptionalism, people sometimes accuse me of *misanthropy*, that is, hatred of humanity. The idea seems to be that I work on this topic because I think that our species is evil and I want to prevent us from realizing our full potential. However, nothing could be further from the truth. I work on this topic because I think that our species can be *good*, and because I want to *support* us in realizing our full potential, by working to build the conditions necessary for higher levels of altruism and flourishing in the future. In my view, it would be misanthropic to *deny* that we have this potential.

Of course, I can see why one might think that human flourishing requires human exceptionalism. Yes, David Bennett might have died shortly after receiving a heart from a pig. But for all he knew at the time, he might have lived longer, and for all we know right now, many humans will have longer and better lives due to xenotransplantation in the future. And of course, many humans still rely on the use of nonhumans for food, clothing, medicine, and other resources that support our survival, too. Thus, there is a limit to how much altruism we can—and should—achieve and sustain beyond our species at present.

How should we think about these tensions between humans and nonhumans? I have argued that we should seek co-beneficial policies for everyone where possible and prioritize thoughtfully where necessary—where this means making at least some sacrifices for nonhumans while still meeting our own basic needs. I have also argued that striking this balance requires pursuing a just transition, by gradually phasing down oppressive practices, gradually phasing up alternatives, and supporting everyone who currently relies on oppressive practices for food, medicine, income, or other basic goods as much as possible along the way.

How should we think about emerging practices like xenotransplantation in this context? Should we phase them up in the short term and then phase them back down again in the long run? Or should we simply not phase them up at all? While I personally lean in the latter direction, I will not insist on that here. However, I will insist on at least this much: We should seek to build a just future for humans and nonhumans alike. And since there is a path dependence to how history unfolds, we should prioritize developing new practices that bring us closer to this just future over developing new practices that bring us even further away from it.

CONCLUSION

Think cosmically, act globally

THE YEAR IS 2100. HUMANS share the world with many nonhuman animals, as well as with silicon-based equivalents of human and nonhuman animals. Carbon-based and silicon-based humans also maintain simulations that contain *simulated* human and nonhuman animals. These beings—carbon-based and silicon-based, physical and simulated—need to decide how to treat everyone when making decisions. For instance, they need to decide whether to prioritize beings made out of one kind of material or beings made out of another, and whether to prioritize beings who exist in one reality or beings who exist in another.

Suppose that you could be reincarnated as any one of these beings. Suppose further that you know everything about this future with one exception: You have no idea which of these beings will be *you*. How would you want these beings to treat each other in this situation? For example, would you want carbon-based and silicon-based humans to give each other equal weight, or would you instead want each population to give extra weight to themselves? Similarly, would you want physical and simulated humans to give each other equal weight, or would you again want each population to give extra weight to themselves?

The year is now 2200. Humans now share the world with new kinds of advanced AI systems whose cognitive and sensory capacities vastly exceed our own. In this world, the difference in cognitive and sensory complexity between these advanced AI systems and humans is about the same as the difference in cognitive and sensory complexity between humans and ants. As a result, the situation is no longer that carbon-based and silicon-based humans need to work together to answer moral, legal, and political questions. The situation is instead that advanced AI systems need to answer these questions for everyone.

Suppose that you could again be reincarnated as any one of these beings, and suppose that you again know everything about this future except which one of these beings will be *you*. How would you want the advanced AI systems to treat humans? For example, would you want them to give us the same amount of weight that they give themselves, despite the fact that we appear less likely to matter *and* likely to matter less from their perspectives? Or would you want them to give us less weight than they give themselves (or, perhaps, no weight at all), in light of the perceived differences between our capacities and theirs?

These cases are adapted from the *original position* thought experiment popularized by John Rawls. In his famous book *A Theory of Justice*, Rawls asks us to imagine that we step behind a *veil of ignorance* where we know everything about society except our own identity, and he asks us to select principles of justice from this position. He then argues that we would select two general principles: a liberty principle according to which everyone should have as much liberty as possible, and a difference principle according to which inequalities are acceptable only when they benefit the worst-off members of society.

While many philosophers dispute these principles, they still agree that the original position is a powerful thought experiment. Like the golden rule ("Do unto others as you would have others do unto you"), the original position might not tell us what to do directly. But by asking us how we would feel if the shoe was on the other foot, it can help us to summon at least *a bit* more impartiality when deciding what to do. Of course, there is a limit to how much impartiality we can summon through a simple thought experiment. But taking a moment to empathize with others before making decisions that affect them can still help.

Rawls spent most of his career writing about justice for human societies with moderate levels of pluralism, scarcity, and inequality, since cooperation is clearly both necessary and possible in such societies. However, toward the end of his career, Rawls extended this project to the global context, and following his death, other scholars have extended it to the multispecies context, too. As Rawls anticipated, these extensions are difficult because cooperation is less clearly possible in such societies. But these extensions are important all the same, and fortunately, scholars are making steady progress on them.

While relatively few scholars have extended this project to the multisubstrate context, this extension is important as well for at least two reasons. First, as in other contexts, if we aspire to treat silicon-based beings as we would hope to be treated in similar circumstances, then we can summon at least *a bit* more empathy and impartiality during our time in power. Of course, this is not to say that we would, or should, accept a maximally inclusive or egalitarian view about the moral circle or about principles of justice. But we might at least conclude that we should strike more of a balance than we do now.

Second, if and when silicon-based beings become more powerful than carbon-based beings, then how we treat them during our time in power might shape how they treat us during their time in power. Thus, this thought experiment is not merely an invitation for us to think *altruistically* about how *we* should treat *them*. It is also an invitation for us to think *selfishly* about how *they* should treat *us*, while we still have time to shape their values. Again, this is not to say that we would, or should, accept a maximally inclusive or egalitarian view. But we might again at least conclude that we should strike more of a balance.

In this respect, part of the beauty and horror of the current trajectory of AI is that it might literally create a kind of heaven or hell for our species. If we follow a virtuous path and treat nonhumans well, then we might eventually find that, having internalized similar values, nonhumans treat us the same way. If, instead, we follow a vicious path and treat nonhumans badly, then we might eventually find that, having internalized similar values, nonhumans treat us the same way. Either way, the future for our species could be either wonderful or terrible, depending in part on how we treat vulnerable others now.

To emphasize, these thought experiments might not compel any particular conclusions directly. But they can still improve our moral reasoning by reminding us that our intuitions about justice depend in part on whether we imagine ourselves in positions of power or vulnerability. If we aspire to a kind of reflective equilibrium between the norms that we think humans should accept now, during our time in power, and the norms that we think AI systems should accept later, during their (potential) time in power, then we might be able to make ourselves less biased now *and* make AI systems less biased in the future.

For whatever it may be worth, my own current view (which, to be clear, is still a work in progress) is strongly inclusive and moderately hierarchical. I believe that if there is *any chance at all* that our actions are helping or harming particular beings, then this possibility merits consideration. But I also believe that we can prioritize some beings over others based on how likely they are to matter, how much they matter if they do, how likely our actions are to impact them, and how much our actions impact them if they do—though we should often "round up" these estimates when in doubt to correct for human bias.

But my main conclusions in this book are more general and ecumenical. If there is at least a non-negligible chance that our actions are helping or harming particular beings, then this possibility merits consideration. In the Anthropocene, there is at least a non-negligible chance that our actions are helping or harming a *vast number and wide range* of beings, including many future nonhumans. And while we can prioritize ourselves to the extent that we need to take care of ourselves and work within our limitations, we should also prioritize ourselves less than we do, and if all goes well, we should eventually not prioritize ourselves at all.

WHEN A TOPIC IS this vast, it can be easy to lose our bearings. Yes, our impacts on sextillions of beings merit consideration. But what does that mean for our lives and societies here and now? Of course, the answer to this question depends on a variety of factors. But it still helps to think in general terms about how to deal with this predicament. So, here are several general aspirations that I think might help. Accepting these aspirations would allow each of us to make more thoughtful decisions about what kinds of lives to lead and what kinds of societies to build in the face of so much complexity and uncertainty.

First, we can cultivate humility about moral standing. The history of thinking about moral standing is a history of unacceptable exclusion and hierarchy, and the mistakes of the past are always clearer than the mistakes of the present. However, we can address this problem to an extent by developing new ways of seeing other kinds of beings. For example, if we use probabilistic language ("high, medium, or low confidence") when discussing whether particular beings matter, and if we use subject language ("they/who") to describe individuals who might matter, then we might be more likely to consider these beings.

Second, and relatedly, we can cultivate humility about our impacts. In the Anthropocene, many individual and collective actions have a small chance of producing a large impact, yet we lack the ability to naturally track these impacts. We can once again address this problem to an extent by developing new ways of assessing our impacts. For example, if we use probabilistic language when describing whether particular impacts will occur, and if we use collective language ("what should *we* do, and how can *I* participate?") when assessing our individual actions, then we might be more likely to consider these impacts.

Third, we can accept that we have a duty to help distant others where possible. We might have this duty because we should help others in general, or we might have it because we might be harming others and we should reduce and repair this harm. Either way, we have this duty. To internalize this idea, we can cultivate empathy for what distant others might be going through and build personal habits and shared structures that make it easier for us to address these problems. At the same time, we can recognize that our ability to do this work will always be limited, and that we need to take care of ourselves to do this work well.

Fourth, we can work to improve our impacts on nearby *and* distant others by searching for interventions that can do at least *some* good for at least *some* humans and nonhumans in the short term, while building the knowledge, power, and political will necessary to do *more* good for *more* humans and nonhumans in the long run. That way, we can take care of ourselves and work within our limitations while expanding our capacity for altruism over time. And we can also lay the groundwork for our successors—whether our successors are humans, nonhumans, or some combination of the two—to take a similarly virtuous path.

To accomplish these aims, we might need to update the common aspiration to *think globally and act locally.* This phrase, which arose during social and environmental movements in the 1960s and 1970s, reflects the idea that global change requires local action. We should seek to change the world at least partly indirectly, by pursuing positive local changes that can contribute to positive global changes. After all, many of us can address local issues more effectively and sustainably than we can address global issues. And if enough of us address local issues in these ways, then we can address global issues together.

However, if my arguments in this book are correct, then we might soon need to extend this idea by aspiring to *think cosmically and act globally* in some cases. In the Anthropocene, many beings might matter, and our actions might be affecting them across time, space, biology, *and* materiality. A century ago, it would have been unthinkable that a multinational company could, say, develop a space program that would lead to countless humans, animals, and AI systems existing in other solar systems and future millennia. But some corporate actors might soon need to contemplate these kinds of distant, large-scale impacts.

We might also need to *think cosmically and act locally*—or, perhaps, *think cosmically, then think globally, then act locally*—in some cases. Suppose that you purchase services from a multinational company that engages in space exploration. Now suppose that you learn one day that this company is planning to send potential moral patients to live in other solar systems without considering the welfare risks for these individuals at all. You might or might not continue to purchase services from this company. But at the very least, your potential complicity in this risky activity should be one factor among many in your decision.

We should take extra care when making decisions that might affect our shared conception of the moral circle. For example, when the New York Court of Appeals rejected the idea that Happy the elephant has a right to liberty, they were right to ask how their decision would affect society. As the majority noted, if society recognizes that elephants can have rights, then we need to ask where to draw the line. Can all vertebrates have rights? What about invertebrates? What about plants and fungi? What about bacteria and viruses? What about robots and chatbots? And if even some of these beings can have rights, then how can society still function?

However, while the majority was right to ask these questions, they were wrong to answer them the way that they did. Happy clearly has a life that matters to her, and we clearly have a responsibility to determine whether her state of captivity is unjust. Moreover, many other nonhuman animals are similarly vulnerable, and in the future, many other nonhumans might be vulnerable too. The majority should have recognized that Happy has a right to liberty, and they should have embraced the opportunity to spark a broader conversation about the scope, strength, and content of nonhuman rights in a careful, thoughtful way.

When Google rejected the idea that LaMDA the large language model was sentient in 2022, they might or might not have considered how their decision or their communications would affect society as well. If they did, I imagine that they reasoned as follows: "This model is almost certainly not sentient, and yet one of our own employees sees it as sentient. As these models become more advanced, the risk of false positives will only increase. We should thus state clearly, simply, and emphatically that LaMDA is non-sentient, both to correct this current false positive and to reduce the risk of future false positives."

However, if Google did reason this way, then their reasoning was incomplete. Yes, we might be at risk of treating objects as subjects. But as our history with other animals illustrates, we are also at risk of treating subjects as objects. And even if the risk of digital sentience was negligible in 2022, it was clearly going to increase over time. Google should have anticipated this trajectory, and—like the New York Court of Appeals, like any of us in a similar situation—they should have embraced the opportunity to spark a broader conversation about the nature and value of other minds in a careful, thoughtful way.

Not everybody will have such a clear opportunity to affect our shared conception of the moral circle. But we all participate in creating the shared beliefs, values, practices, traditions, institutions, and infrastructures that govern our shared existence. By including everyone who might matter in our individual conceptions of the moral circle, and by considering everyone we might be affecting when selecting the rules, habits, and structures that govern our individual and collective actions, we can participate in building a more respectful and compassionate society. The more we all do that, the better off everyone will, or at least might, be.

ACKNOWLEDGMENTS

This book is based on my paper "Moral Circle Explosion," which I contributed to *The Oxford Handbook of Animal Ethics*. It adapts material from other recent work as well, including (in order of appearance) "The Moral Problem of Other Minds," in the *Harvard Review of Philosophy*; "The Rebugnant Conclusion," in *Ethics, Policy, & Environment*; "On the Torment of Insect Minds and the Duty Not to Farm Them" (with Jason Schukraft), in Aeon; and "Against Human Exceptionalism," also in Aeon. Thanks to everyone who contributed to this material, including Jason and all the editors and reviewers.

Tom Mayer and Andrew Blitzer, my editors at W. W. Norton, developed this book with me from the ground up and provided excellent feedback on multiple drafts. Many others at W. W. Norton supported the book as well, including Alane Mason and Justin Cahill (Norton Shorts series founders, along with Tom Mayer), Caroline Adams and YJ Wang (editorial assistants), Kyle Radler (series publicist), TK TK (book publicist and marketer), Delaney Adams (production manager), Susan Sanfrey (project editor), Sarah Johnson (copyeditor), Steve Attardo (series jacket design), and Sarahmay Wilkinson (jacket design).

Several other people played a major role in shaping the content of this book as well. Most importantly, Toni Sims provided extensive research support, editorial support, and substantive feedback, both on this book and on other papers and essays from which the book draws. Additionally, Adam Bales, Lucius Caviola, David Chalmers, Colin Marshall, and Stefan Schubert provided very helpful feedback on previous drafts. Thanks to everyone, particularly Toni, for your contributions to this project. For all its faults, the book is much better than it otherwise could have been as a result of your feedback and support.

I presented material from this book at many places over the past two years, including—but not limited to—Complutense University, Pompeu Fabra University, Princeton University, the University of Hamburg, the University of Oxford, the University of Santiago de Compostela, the University of Texas at Austin, the Mimir Center for Long Term Futures Research, and the World Forum on the Future of Democracy, Tech, and Humankind. The Global Priorities Institute at Oxford was particularly supportive, hosting multiple talks and discussions related to this book. Thanks to the organizers and participants in all cases.

Speaking of institutional support, I would never have been able to write this book without my amazing colleagues at New York University, particularly my colleagues in animal studies, environmental studies, bioethics, medical ethics, philosophy, and law. I especially owe a lot to my colleagues at the Center for Environmental and Animal Protection; the Center for Mind, Ethics, and Policy; the Wild Animal Welfare Program; the Guarini Center on Environmental, Energy, and Land Use Law; and the More Than Human Rights Project. Thanks to everyone across these units for advancing my understanding of this topic.

While some colleagues contributed to this book more directly than others, several people deserve special mention for how much they impacted my work during this period in general. In particular, thanks to

Dale Jamieson for your mentorship and support; to Sofia Fogel, Becca Franks, and (again) Toni Sims for working so closely with me at NYU; to Katrina Wyman, Adalene Minelli, and Cleo Verkuijl for working with me on multiple projects related to legal change for nonhumans; and to Kristin Andrews, Jonathan Birch, and Robert Long for working with me on multiple projects related to nonhuman consciousness.

On a personal note, thanks to my partner Lindsey Liberatore for all your love and support over the past two years. Thanks to my parents Eric and Sheryl Sebo, my brother Marc Sebo, and my former partner, dear friend, and dog co-parent Maryse Mitchell-Brody for the same. Finally, and most importantly: This book is dedicated to my dog Smoky Sebellody, who has lived with me and Maryse (now alternating between us, along with our respective partners) for more than a decade. While I reject human exceptionalism, I remain a Smoky exceptionalist. Smoky is the most significant being who will ever live, and I feel grateful to know him.

NOTES ON SOURCES AND QUOTATIONS

INTRODUCTION: THE ELEPHANT AND THE CHATBOT

000 **the Nonhuman Rights Project argued**: Brief for Petitioner-Appellant at 7, Happy ex rel. Nonhuman Rights Project, Inc. v. Breheny, APL-2021-00087 (filed July 2, 2021).

000 **The zoo replied**: Brief for Respondents-Respondents at 7, Happy ex rel. Nonhuman Rights Project, Inc. v. Breheny, APL-2021-00087 (filed August 20, 2021).

000 **the court sided with the zoo**: Nonhuman Rights Project, Inc. v. Breheny, WL 2122141 (New York Court of Appeals 2022).

000 **vulnerable to harm, exploitation, and extermination**: Kristin Andrews et al., *Chimpanzee Rights: The Philosophers' Brief* (Routledge, 2018).

000 **alleging that a chatbot named LaMDA was sentient**: Nitasha Tiku, "The Google Engineer Who Thinks the Company's AI Has Come to Life," *Washington Post*, June 11, 2022.

000 **"would be exactly like death for me"**: Blake Lemoine, "Is LaMDA Sentient? An Interview," *Medium* (blog), June 11, 2022.

000 **violating its confidentiality policy**: Tiku, "Google Engineer."

000 **spokesperson Brian Gabriel noted**: Tiku.

000 **requested legal representation**: Steven Levy, "Blake Lemoine Says Google's LaMDA AI Faces 'Bigotry,'" *Wired*, June 17, 2022.

000 **Google fired Lemoine**: Nico Grant, "Google Fires Engineer Who Claims Its A.I. Is Conscious," *New York Times*, July 23, 2022.

000 ***moral circle expansion***: For early discussion of the idea of circles of moral sympathy: see Adam Smith, *The Theory of Moral Sentiments* (Liberty Classics,

1976). For later, but still early, discussion of the idea of moral circle expansion, see: Peter Singer, *Animal Liberation* (Avon Books, 1975).

000 **Our impact on beings—both present and future—is intensifying**: For discussion of the Anthropocene, see Michael Balter, "Archaeologists Say the 'Anthropocene' Is Here—But It Began Long Ago," *Science*, April 19, 2013, 261–62; Simon L. Lewis and Mark A. Maslin, "Defining the Anthropocene," *Nature* 519, no. 7542 (March 2015): 171–80; Jeff Sebo, *Saving Animals, Saving Ourselves: Why Animals Matter for Pandemics, Climate Change, and Other Catastrophes* (Oxford University Press, 2022); Jan Zalasiewicz et al., "The Anthropocene: A New Epoch of Geological Time?," *Philosophical Transactions of the Royal Society A: Mathematical, Physical and Engineering Sciences* 369, no. 1938 (2011): 835–41.

000 ***creating* beings, including both animals and AI systems**: For example, see discussions of chimeras: Insoo Hyun, "What's Wrong with Human/Nonhuman Chimera Research?," *PLOS Biology* 14, no. 8 (2016). See also discussions of artificial intelligence: David J. Gunkel, "The Other Question: Can and Should Robots Have Rights?," *Ethics and Information Technology* 20, no. 2 (2018): 87–99.

CHAPTER 1: MORAL STATUS

000 **Universal Declaration of Human Rights**: United Nations, "Universal Declaration of Human Rights," 1948.

000 **they see humanity as valuable in itself**: Bernard Williams, "The Human Prejudice," in *Philosophy as a Humanistic Discipline*, ed. A. W. Moore (Princeton University Press, 2006).

000 **they see humanity as the source of other valuable features**: Immanuel Kant, *Groundwork for the Metaphysics of Morals*, trans. Mary J. Gregor (1785; repr., Cambridge University Press, 1996).

000 **we should achieve justice for our own species**: Abeba Birhane and Jelle van Dijk, "Robot Rights? Let's Talk about Human Welfare Instead," *Proceedings of the AAAI/ACM Conference on AI, Ethics, and Society* (Association for Computing Machinery, 2020), 207–13.

000 **inspired by a similar case from Dale Jamieson**: The discussion of surviving members of the *Homo neanderthalensis* species is adapted from Dale Jamieson's *Ethics and the Environment*. Jamieson also considers a species called *Homo floresiensis*, which purportedly evolved on the island of Flores, and he imagines a hominid extraterrestrial species called "Trafalmadoreans." See: Dale Jamie-

son, *Ethics and the Environment: An Introduction* (Cambridge University Press, 2008), 107–12.

000 **moral standing**: Dale Jamieson, "Humans and Other Animals," in *Ethics and the Environment*, 107–12; Jeff Sebo, "Moral Circle Explosion," in *The Oxford Handbook of Normative Ethics*, ed. David Copp, Connie Rosati, and Tina Rulli (Oxford University Press, forthcoming); Mary Anne Warren, "The Concept of Moral Status," in *Moral Status: Obligations to Persons and Other Living Things* (Oxford University Press, 2000), 3–23.

000 **Philosophers use the terms *moral patient* and *moral agent***: Evelyn Pluhar, "Moral Agents and Moral Patients," *Between the Species* 4, no. 1 (1988); Tom Regan, *The Case for Animal Rights* (University of California Press, 1983).

000 **different duties to different moral patients**: The content and strength of the moral duties that are typical for members of a certain species may depend on the capacities and kinds of relationships that are typical for members of that species. For a comprehensive view that links moral rights and responsibilities with species-typical capabilities, see Martha C. Nussbaum, *Justice for Animals: Our Collective Responsibility* (Simon & Schuster, 2023). For further discussion of the role of capacities in moral duty, see: Gary Watson, "Two Faces of Responsibility," *Philosophical Topics* 24, no. 2 (1996): 227–48; Susan Wolf, "Sanity and the Metaphysics of Responsibility," in *Responsibility, Character, and the Emotions: New Essays in Moral Psychology*, ed. Ferdinand David Schoeman (Cambridge University Press, 1987), 46–62.

For further discussion of the role of relationships in moral duty, see: Mark Coeckelbergh, "Robot Rights? Towards a Social-Relational Justification of Moral Consideration," *Ethics and Information Technology* 12, no. 3 (2010): 209–21; Katrina Hutchison, Catriona Mackenzie, and Marina Oshana, eds., *Social Dimensions of Moral Responsibility* (Oxford University Press, 2018); R. Jay Wallace, *The Moral Nexus* (Princeton University Press, 2019).

000 ***the welfare principle***: Philosophers who accept the principle that we can have moral duties toward any individuals who have the capacity for welfare include Kenneth E. Goodpaster, Tom Regan, and Peter Singer. See: Kenneth E. Goodpaster, "On Being Morally Considerable," *Journal of Philosophy* 75, no. 6 (1978): 308–25; Regan, *Case for Animal Rights*; Singer, *Animal Liberation*.

000 **The welfare principle implies that we should reject *rationalism***: Jeremy Bentham was one of the earliest philosophers who rejected rationalism in favor of the welfare principle. More modern proponents of the welfare principle include Peter Singer and Wayne Sumner. See: Jeremy Bentham, *An Introduction to the Principles of Morals and Legislation*, vol. 45 (1789; repr., Dover

Publications, 2007); Peter Singer, "All Animals Are Equal," in *Applied Ethics: Critical Concepts in Philosophy*, ed. Ruth Chadwick and Doris Schroeder (Routledge, 1986): 4:51–79; L. Wayne Sumner, *Welfare, Happiness, and Ethics* (Clarendon Press, 1996).

000 **we should reject *speciesism***: The term "speciesism" was coined by Richard Ryder in a 1970 privately distributed leaflet. It was later popularized by Peter Singer's book *Animal Liberation*. See: Richard D. Ryder, "Speciesism Again: The Original Leaflet," *Critical Society* 2, no. 1 (2010); Singer, *Animal Liberation*.

000 **avoid kicking them unnecessarily**: Peter Singer, *Practical Ethics*, 2nd ed. (Cambridge University Press, 1993), 58.

000 **sentient beings are welfare subjects**: Christine Korsgaard, *Fellow Creatures: Our Obligations to the Other Animals* (Oxford University Press, 2018); Regan, *Case for Animal Rights*; Singer, *Animal Liberation*.

000 **for Dara to count as sentient in the relevant sense**: Gwen Bradford, "Consciousness and Welfare Subjectivity," *Noûs* 57 (2022); Neil Levy and Julian Savulescu, "Moral Significance of Phenomenal Consciousness," in *Coma Science: Clinical and Ethical Implications*, ed. Steven Laureys, Nicholas D. Schiff, and Adrian M. Owen, *Progress in Brain Research* 177 (2009), 361–70; Eden Lin, "The Experience Requirement on Well-Being," *Philosophical Studies* 178, no. 3 (2021): 867–86.

000 **bad for you otherwise**: Nicolas Delon et al., "Consider the Agent in the Arthropod," *Animal Sentience* 5, no. 29 (2020); Shelly Kagan, *How to Count Animals, More or Less* (Oxford University Press, 2019), 16–30; Ali Ladak, "What Would Qualify an Artificial Intelligence for Moral Standing?," *AI and Ethics* (2023).

000 **is this kind of subjective awareness necessary for Dara to count**: Kenneth E. Himma, "Artificial Agency, Consciousness, and the Criteria for Moral Agency: What Properties Must an Artificial Agent Have to Be a Moral Agent?," *Ethics and Information Technology* 11 (2009): 19–29.

000 **living beings are welfare subjects**: Goodpaster, "On Being Morally Considerable"; Paul W. Taylor, *Respect for Nature*, 25th anniversary ed. (Princeton University Press, 2011); Leena Vilkka, "Biocentrism," in *The Intrinsic Value of Nature*, Value Inquiry Book Series 59 (Brill, 1997), 53–70.

000 **Are a carbon-based substrate and an evolutionary origin necessary**: Fred Feldman, "Vitalist Theories of Life," in *Confrontations with the Reaper* (Oxford University Press, 1994), 51–53.

000 **the *moral weight* of lives**: Bob Fischer, "An Introduction to the Moral Weight Project," Rethink Priorities, October 31, 2022; Kagan, *How to Count Animals*.

000 ***the principle of equal consideration of interests***: Singer, *Animal Liberation*. Peter Singer argues for the equal consideration of interests regardless of species. He does not discuss the equal consideration of interests regardless of substrate.

000 **the elephant has a stronger interest in avoiding suffering**: Peter Singer, "All Animals Are Equal," in *Animal Liberation*, 40th anniversary ed. (Open Road Media, 2015), 26–57.

000 **neither kind of life is better as a general matter**: Korsgaard, *Fellow Creatures*.

000 **happiness, suffering, and other such welfare states**: Fischer, "Introduction to the Moral Weight Project"; Kagan, *How to Count Animals*.

000 **how exactly are welfare capacities related to brain complexity?**: Adam Shriver, "What Neuron Counts Can and Can't Tell Us about Moral Weight," Moral Weight Project, Rethink Priorities, 2022.

000 **proxy for our welfare capacity**: Jason Schukraft, "The Subjective Experience of Time: Welfare Implications," Moral Weight Project, Rethink Priorities, July 27, 2020.

000 **an individual elephant carries more weight than *any number of* ants**: J. Paul Kelleher, "Relevance and Non-Consequentialist Aggregation," *Utilitas* 26, no. 4 (2014): 385–408; George Edward Moore, *Principia Ethica* (1903; repr., Cambridge University Press, 1993), 70–80; John M. Taurek, "Should the Numbers Count?," *Philosophy & Public Affairs* (1977): 293–316.

000 **large number of ants really *can* carry more weight than a single elephant**: John Broome, *Weighing Lives* (Oxford Academic, 2005); Joe Horton, "Always Aggregate," *Philosophy and Public Affairs* 46, no. 2 (2018): 160–74; Alastair Norcross, "Comparing Harms: Headaches and Human Lives," *Philosophy and Public Affairs* 26, no. 2 (1997): 135–67.

000 ***Reasons and Persons***: Derek Parfit, *Reasons and Persons* (Oxford University Press, 1984).

000 **the repugnant conclusion**: Parfit, *Reasons and Persons*; Derek Parfit, "Overpopulation and the Quality of Life," in *The Repugnant Conclusion: Essays on Population Ethics* (Springer Netherlands, 2004), 7–22; Derek Parfit, "Can We Avoid the Repugnant Conclusion?," *Theoria* 82, no. 2 (2016): 110–27.

000 **the rebugnant conclusion**: Jeff Sebo, "Animals, Creation Ethics, and Population Ethics," in *Saving Animals, Saving Ourselves*, 166–90; Jeff Sebo, "The Rebugnant Conclusion: Utilitarianism, Insects, Microbes, and AI Systems," *Ethics, Policy & Environment* 26 (2023): 1–16.

000 **Intersubstrate welfare comparisons**: Bob Fischer and Jeff Sebo, "Intersubstrate Welfare Comparisons: Important, Difficult, and Potentially Tractable," *Utilitas* 36, no. 1 (2023): 1–14.

000 **France, Quebec, Colombia, Mexico City, and other governments**: Andrews et al., *Chimpanzee Rights*, 119–20.

000 **legal authorities now recognize some animals as rights-holders**: Animal Law & Policy Program, "Constitutional Court of Ecuador Recognizes Animal Rights in Landmark Ruling," Harvard Law School, media release, March 23, 2022; Kevin Schneider, "The Case of Laxmi the Elephant and Animal Rights in India," Nonhuman Rights Project, August 14, 2020; Taylor Gulatsi, "Legal Personality for Animals in India and Pakistan," *In Custodia Legis* (blog), Library of Congress, August 30, 2023.

CHAPTER 2: MORAL THEORY

000 **a more just and sustainable future**: United Nations, "Sustainable Development Goals," https://sdgs.un.org/goals.

000 **2013 report commissioned by the UN**: High-level Panel of Eminent Persons on the Post-2015 Development Agenda, *A New Global Partnership: Eradicate Poverty and Transform Economies through Sustainable Development*, United Nations, 2013.

000 **2021 UN initiative**: Office of the Secretary-General, *Our Common Agenda: Report of the Secretary-General*, United Nations, 2021.

000 **less than 1% of its annual gross domestic product (GDP)**: In 2019, the United States spent $39.2 billion on foreign aid. See: George Ingram, "What Every American Should Know about US Foreign Aid," Brookings Institution, October 15, 2019.

000 **a similarly paltry amount on efforts**: In 2022, the United States spent $36.6 billion directly on infrastructure and provided $94.5 billion to states for infrastructure. See: "2023 State of the Union: US Transportation & Infrastructure," USAFacts, accessed April 19, 2024.

000 **nations recognize future generations in their constitutions**: In the 1960s, fewer than ten nations referenced future generations in their constitutions. As of 2021, 81 out of the 196 UN member states reference future generations in their constitutions. Renan Araújo and Leonie Koessler, "The Rise of the Constitutional Protection of Future Generations," LPP Working Paper No. 7-2021, SSRN, last revised November 8, 2022.

000 **not yet led to significant policy change**: Araújo and Koessler, "Rise of the Constitutional Protection."

000 **rights of individual animals at all**: Cleo Verkuijl et al., "Mainstreaming

Animal Welfare in Sustainable Development: A Policy Agenda," Stockholm Environmental Institute, Stockholm+50 Background Paper Series, 2022.

000 **difficult to make good decisions in everyday life through theoretical deliberation alone**: Richard Hare, *Moral Thinking: Its Levels, Method, and Point* (Clarendon Press, 1981); Tyler John and Jeff Sebo, "Consequentialism and Nonhuman Animals," in *The Oxford Handbook of Consequentialism*, ed. Douglas W. Portmore (Oxford University Press, 2020); Eric Wiland, "How Indirect Can Indirect Utilitarianism Be?," *Philosophy and Phenomenological Research* 74, no. 2 (2007): 275–301.

000 ***consequentialism* holds that morality is primarily about *consequences***: Historically important consequentialists include Jeremy Bentham, John Stuart Mill, and Henry Sidgwick: Bentham, *Introduction to the Principles of Morals and Legislation*; J. S. Mill, *Utilitarianism*, ed. Roger Crisp (1861; repr., Oxford University Press, 1998); H. Sidgwick, *The Methods of Ethics*, 7th ed. (1874; repr., Macmillan, 1907).

000 ***rights theory* holds that morality is primarily about rights**: Prominent rights theorists include Immanuel Kant, Christine Korsgaard, and Tom Regan: Kant, *Groundwork for the Metaphysics of Morals*; Korsgaard, *Fellow Creatures*; Regan, *Case for Animal Rights*.

000 ***virtue ethics* and *care ethics* hold that morality**: For more on virtue theory, see the work of Aristotle, Rosalind Hursthouse, and Liezl L. Van Zyl: Aristotle, *Nicomachean Ethics*, ed. and trans. Robert C. Bartlett and Susan D. Collins (University of Chicago Press, 2012); Rosalind Hursthouse, *On Virtue Ethics* (Oxford University Press, 1999); Liezl L. Van Zyl, *Virtue Ethics: A Contemporary Introduction* (Routledge, 2018).

For more on care theory, see the work of Stephanie Collins and Nel Noddings: Stephanie Collins, "Care Ethics: The Four Key Claims," in *Moral Reasoning*, ed. David R. Morrow (Oxford University Press, 2017); Nel Noddings, *Caring: A Feminine Approach to Ethics and Moral Education* (University of California Press, 1986); Carol J. Adams and Lori Gruen, eds., *Ecofeminism: Feminist Intersections with Other Animals and the Earth* (Bloomsbury, 2021); and Josephine Donovan and Carol J. Adams, eds., *The Feminist Care Tradition in Animal Ethics: A Reader* (Columbia University Press, 2007).

000 **he starts dumping his waste in the river**: Sadly, David's new solution is not entirely unrealistic; chemical companies have a history of dumping pollution in places like Toms River, New Jersey. See: Dan Fagin, *Toms River* (Island Press, 2015).

000 **the other is thousands of miles away from me**: Peter Singer, "Famine, Affluence, and Morality," *Philosophy and Public Affairs* 1, no. 3 (1972): 229–43.

000 **Entangled lives require entangled ethics**: Lori Gruen, *Entangled Empathy: An Alternative Ethic for Our Relationships with Animals* (Lantern Books, 2015).

000 **we take on special duties to each other, too**: Baruch Brody, "Defending Animal Research: An International Perspective," in *The Ethics of Animal Research: Exploring the Controversy*, ed. Jeremy Garrett (MIT Press, 2012).

000 **than they would otherwise experience**: This account of harm is sometimes called the counterfactual comparative account of harm. For more on the nature of harm, see: Neil Feit, *Bad Things: The Nature and Normative Role of Harm* (Oxford University Press, 2023); Justin Klocksiem, "A Defense of the Counterfactual Comparative Account of Harm," *American Philosophical Quarterly* 49, no. 4 (2012): 285–300; Duncan Purves, "Harming As Making Worse Off," *Philosophical Studies* 176, no. 10 (2019): 2629–56.

000 ***the subject-relative view***: The view that you harm someone only when you make them worse off than they otherwise would be is sometimes called the person-specific or person-affecting view. I prefer the term "subject-relative" because it avoids questions about personhood. For more on this view, see: Gustaf Arrhenius, "The Person-Affecting Restriction, Comparativism, and the Moral Status of Potential People," *Ethical Perspectives* 10, no. 3 (2003): 185–95; Christopher J. G. Meacham, "Person-Affecting Views and Saturating Counterpart Relations," *Philosophical Studies* 158, no. 2 (2012): 257–87.

000 ***the subject-neutral view***: Nils Holtug, "Who Cares about Identity?," in *Harming Future Persons*, ed. M. A. Roberts and D. T. Wasserman (Springer, 2009), 71–92.

000 **that intuition might or might not be correct**: Derek Parfit, "The Non-Identity Problem," in *Reasons and Persons*, 351–80.

000 **actions and policies that change the identities of future subjects**: Parfit, "Non-Identity Problem."

000 ***the swan identity problem***: Sebo, "Animals, Creation Ethics, and Population Ethics."

000 **animal welfare and sustainable development**: Amelia Linn and Sebastian Osborn, "One Year Ago, the United Nations Adopted a Landmark Animal Welfare Resolution," *Mercy for Animals* (blog), March 9, 2023.

CHAPTER 3: IF YOU MIGHT MATTER, WE SHOULD ASSUME YOU DO

000 **"Consider the Lobster"**: Wallace's article is also available in a collection of his essays. David Foster Wallace, *Consider the Lobster: And Other Essays* (Little, Brown, 2004).

000 **the moral problem of other minds**: Jeff Sebo, "The Moral Problem of Other Minds," *Harvard Review of Philosophy* 25 (2018): 51–70. The introduction of this chapter is adapted from the introduction of this paper.

000 **this risk would have at least merited consideration, despite being unlikely**: In fact, Oppenheimer *was* concerned about the risk of accidentally igniting the Earth's atmosphere, though his colleague, physicist Hans Bethe, convinced him that there was a near-zero chance of such a catastrophe. Kai Bird and Martin J Sherwin, *American Prometheus* (Vintage Books, 2006), 183.

000 **our generation is similarly fallible**: For a discussion of our evolving views on animal consciousness, see: Kristin Andrews, *The Animal Mind: An Introduction to the Philosophy of Animal Cognition* (Routledge, 2015).

000 **between one in ten thousand and one in ten quadrillion**: For a summary of different philosophers' suggested risk thresholds, see: Bradley Monton, "How to Avoid Maximizing Expected Utility," *Philosophers' Imprint* 19, no. 18 (2019): 1–25.

000 **the case for the no threshold view**: Sebo, "Rebugnant Conclusion"; Hayden Wilkinson, "In Defense of Fanaticism," *Ethics* 132, no. 2 (2022).

000 **the case for the threshold view**: Monton, "How to Avoid Maximizing."

000 **Pascal's mugging, in honor of mathematician Blaise Pascal**: Eliezer Yudkowsky, "Cognitive Biases Potentially Affecting Judgment of Global Risks," in *Global Catastrophic Risks*, ed. Nick Bostrom and Milan M. Ćirković (Oxford University Press, 2008), 91–119.

000 **Jonathan Birch and colleagues published a detailed review**: Jonathan Birch et al., *Review of the Evidence of Sentience in Cephalopod Molluscs and Decapod Crustaceans* (London School of Economics and Political Science, 2021).

000 **local anesthetics or analgesics**: Birch et al., *Review of the Evidence.*

000 **a one in ten chance of being sentient**: Birch et al.

000 **Bill in 2021 to include invertebrates like crabs, lobsters, and octopuses**: "Lobsters, Octopus and Crabs Recognised as Sentient Beings," Gov.UK, 2021.

000 **remained roughly stable in the UK for the past decade**: According to the Food and Agriculture Organization of the United Nations (FAOSTAT), the food supply of crustaceans in the UK has been between 3.17 and 3.79 kg per capita per year since 2010.

CHAPTER 4: MANY BEINGS MIGHT MATTER

000 **In January of 2023**: The introduction of this chapter is adapted from material originally written by Jason Schukraft for our co-authored essay on insect farming: Jeff Sebo and Jason Schukraft, "On the Torment of Insect Minds and Our Moral Duty Not to Farm Them," Aeon, July 27, 2021.

000 **broke ground on a production site**: "InnovaFeed Breaks Ground on Decatur Facility," press release, City of Decatur, Illinois, January 10, 2023.

000 **sixty thousand metric tons of fly larvae protein each year**: Clint Rainey, "This Thriving French Insect Farm Is Opening an Even Bigger Bug Factory in Illinois," Fast Company, March 19, 2024.

000 **three billion metric tons of protein per year**: "Global Edible Insects Market Report 2022–2030: Environmental Benefits of Edible Insects Consumption & Rising Demand for Insect Protein in the Animal Feed Industry," GlobeNewswire Newsroom, June 15, 2022.

000 **no laws**: Matt Reynolds, "Insect Farming Is Booming. But Is It Cruel?," *Wired*, March 16, 2023.

000 **injury, disease, conflict, and cannibalism**: M. Barrett et al., "Welfare Considerations for Farming Black Soldier Flies, Hermetia Illucens (Diptera: Stratiomyidae)," *Journal of Insects as Food and Feed* 9, no. 2 (2023): 119–48; Helen Lambert, "Insect Farming and Sustainable Food Systems," Eurogroup for Animals, 2022.

000 **who sometimes eat them alive**: Barrett et al., "Farming Black Soldier Flies."

000 **agricultural pesticides**: H. J. B. Howe, *Improving Pest Management for Wild Insect Welfare*, Wild Animal Initiative, December 2019.

000 **some to contract and others to expand**: Jeff Sebo, "Animals, Pandemics, and Climate Change," in *Saving Animals, Saving Ourselves*, 40–65.

000 **cephalopod mollusks, decapod crustaceans, and insects**: Kristen Andrews et al., "The New York Declaration on Animal Consciousness," 2024, nydeclaration.com.

000 **arguments that Birch and I make elsewhere**: Jonathan Birch, *The Edge of Sentience: Risk and Precaution in Humans, Other Animals, and AI* (Oxford University Press, 2024), also available online (open access); Jonathan Birch,

"Animal Sentience and the Precautionary Principle," *Animal Sentience* 16, no. 1 (2017); Sebo, "Moral Problem of Other Minds."

000 **our responses to these risks**: Andrews et al., "New York Declaration on Animal Consciousness."

000 **bacteria and archaea**: Samantha Fowler, Rebecca Roush, and James Wise, "Unit 4: Evolution and the Diversity of Life," in *Concepts of Biology* (OpenStax, 2013).

000 **one central brain and another, smaller, connected brain in each arm**: Peter Godfrey-Smith, *Other Minds: The Octopus, the Sea, and the Deep Origins of Consciousness* (Farrar, Straus and Giroux, 2016).

000 **and in the two octopuses as a pair**: Luke Roelofs and Jeff Sebo, "Overlapping Minds and the Hedonic Calculus," *Philosophical Studies* (forthcoming).

000 **how is this any different?**: Peter Carruthers, "The Problem of Other Minds," in *The Nature of the Mind: An Introduction* (Routledge, 2003), 6–39.

000 **a basic property of all matter**: This view is called "panpsychism." See: Godehard Bruntrup and Ludwig Jaskolla, eds., *Panpsychism: Contemporary Perspectives* (Oxford University Press, 2016).

000 **beliefs and desires**: For more on these relatively undemanding theories of consciousness see: Anil K. Seth and Tim Bayne, "Theories of Consciousness," *Nature Reviews Neuroscience* 23, no. 7 (2022): 439–52; Jeff Sebo and Robert Long, "Moral Consideration for AI Systems by 2030," *AI and Ethics* (forthcoming).

000 **higher-order thoughts**: Richard Brown, Hakwan Lau, and Joseph E. LeDoux, "Understanding the Higher-Order Approach to Consciousness," *Trends in Cognitive Sciences* 23, no. 9 (2019): 754–68.

000 **chemical and electric signals**: Ned Block, "Comparing the Major Theories of Consciousness," in *The Cognitive Neurosciences*, ed. M. S. Gazzaniga, E. Bizzi, and L. M. Chalupa (MIT Press, 2009), 1111–22.

000 **perception, attention, learning, memory, self-awareness, and decision-making**: Patrick Butlin et al., "Consciousness in Artificial Intelligence: Insights from the Science of Consciousness," arXiv, August 22, 2023.

000 **very little evidence about most insect species**: For discussion, see: Daniela Waldhorn et al., "Invertebrate Sentience Table," Invertebrate Welfare Cause Profile, Rethink Priorities, 2019.

000 **a common behavior in animals who can feel pain**: E. Leadbeater and L. Chittka, "Social Learning in Insects: From Miniature Brains to Consensus Building," *Current Biology* 17 (2007): R703–13; Julia Groening, Dustin Venini, and Mandyam Srinivasan, "In Search of Evidence for the Experi-

ence of Pain in Honeybees: A Self-Administration Study," *Scientific Reports* 7 (2017).

000 **nematodes . . . react appropriately to stimuli**: For information on nematode reactions, see: Andrew D. Goldsmith et al., "Developmental Control of Lateralized Neuron Size in the Nematode *Caenorhabditis elegans*," *Neural Development* 5 (2010): 1–17.

000 **plants . . . react appropriately to stimuli**: For information on plant behavior, see: Paco Calvo and Natalie Lawrence, *Planta Sapiens: Unmasking Plant Intelligence* (Hachette, 2022); Peter Wohlleben, *The Hidden Life of Trees: What They Feel, How They Communicate; Discoveries from a Secret World*, vol. 1 (Greystone Books, 2016).

000 **limited knowledge about how they perform this task**: Aleksandre Asatiani et al., "Challenges of Explaining the Behavior of Black-Box AI Systems," *MIS Quarterly Executive* 19, no. 4 (2020): 259–78.

000 **possess these indicators in the future**: Butlin et al., "Consciousness in Artificial Intelligence."

000 **on a 2020 survey of philosophers**: David Bourget and David Chalmers, "Philosophers on Philosophy: The 2020 Philpapers Survey," *Philosophers' Imprint* 23, no. 1 (2023).

000 **quadrillions, if not quintillions, of animals alive**: Hollis J. B. Howe, "Improving Pest Management for Wild Insect Welfare," Wild Animal Initiative, December 2019, 5.

000 **"A Defense of Abortion," Judith Jarvis Thomson**: Judith Jarvis Thomson, "A Defense of Abortion," in *Ethics: Contemporary Readings*, ed. Harry Gensler, Earl Spurgin, and James Swindal (Routledge, 2004), 267–74.

000 **difficult to grasp**: This conclusion draws from material originally written by Jason Schukraft for our co-authored essay on insect farming: Sebo and Schukraft, "Torment of Insect Minds."

000 **roughly forty-five to fifty trillion insects**: Lambert, "Insect Farming and Sustainable Food Systems."

000 **more than three thousand times**: In our essay for Aeon, Jason Schukraft and I wrote that 60,000 tons of insects, lined up, would stretch to the moon and back 25 times. I used the same estimates for this calculation. Sebo and Schukraft, "Torment of Insect Minds."

000 **14.5% of all human-caused greenhouse gas emissions**: P. J. Gerber et al., "Tackling Climate Change through Livestock: A Global Assessment of Emissions and Mitigation Opportunities," Food and Agriculture Organization of the United Nations (FAO), 2013.

000 **for insect feed**: Lynn Fantom, "Insect Farms Are Scaling Up—and Crossing the Atlantic—in a Play for Sustainable Protein," Civil Eats, December 20, 2022.

000 **insect protein for fish feed**: "Innovafeed and Cargill Extend Their Partnership to Bring Healthy Novel Ingredients to Aquafarmers," press release, Cargill, June 30, 2022.

000 **use insects as feed**: Abraham Rowe, "Insects Raised for Food and Feed: Global Scale, Practices, and Policy," Rethink Priorities, 2020.

CHAPTER 5: IF WE MIGHT BE AFFECTING YOU, WE SHOULD ASSUME WE ARE

000 **They were right**: Dylan Matthews, "40 Years Ago Today, One Man Saved Us from World-Ending Nuclear War," Vox, September 26, 2023.

000 **allowing the rabbit to die**: This example is adapted from Chapter 1 of my book *Saving Animals, Saving Ourselves*, which in turn is adapted from Peter Singer's article "Famine, Affluence, and Morality."

000 **a multi-issue, "both-and" lens**: Jeff Sebo, "Multi-Issue Food Activism," in *The Oxford Handbook of Food Ethics*, ed. Anne Barnhill, Mark Budolfson, and Tyler Doggett (Oxford University Press, 2018), 399–426.

000 **the assassins can be collectively responsible**: Christopher Kutz, *Complicity: Ethics and Law for a Collective Age* (Cambridge University Press, 2000).

000 ***rule consequentialism***: Brad Hooker, "Rule-Consequentialism," *Mind* 99, no. 393 (1990).

000 **the *universal law principle***: Immanuel Kant famously argued that one way to assess the rightness of an action is to consider whether we can will, or endorse, the maxim of that action as a universal law. See: Kant, *Groundwork for the metaphysics of morals*, 4:421.

000 **these insects clearly merit consideration in this context**: Gary David O'Brien, "Directed Panspermia, Wild Animal Suffering, and the Ethics of World-Creation," *Journal of Applied Philosophy* 39, no. 1 (2022): 87–102.

000 **the many nonhuman victims of our food system**: Amelia Linn, "Climate-Friendly Diets: How Cities Can Cut Emissions and Improve Public Health," Guarini Center, NYU School of Law, 2018; "Mayor de Blasio Announces Citywide Meatless Mondays," City of New York, March 11, 2019, NYC.gov.

000 **deadliest European invasion since World War II**: Serhii Plokhy, *The Russo-Ukrainian War* (Penguin, 2023).

000 **threatening a nuclear attack in Ukraine**: David E. Sanger, "New Nuclear

Threats from Putin, Timed for a Moment of Anxiety," *New York Times*, February 29, 2024.

000 **unacceptably high**: "Risk of Nuclear Weapons Use Higher Than at Any Time since Cold War, Disarmament Affairs Chief Warns Security Council," Meetings Coverage and Press Releases, United Nations, March 31, 2023.

000 **might soon join the nuclear club as well**: Bastian Herre et al., "Nuclear Weapons," Our World in Data, March 18, 2024.

000 **facilitating the creation of harmful pathogens**: Jonas B. Sandbrink, "Artificial Intelligence and Biological Misuse: Differentiating Risks of Language Models and Biological Design Tools," preprint, arXiv:2306.13952 (2023); Toby Ord, "Future Risks," in *The Precipice: Existential Risks and the Future of Humanity* (Hachette, 2020), 121–58.

000 **contributing to industrialization**: Toby Ord, "Anthropogenic Risks," in *The Precipice*, 89–113.

000 **contributing to surveillance and military technology**: Steven Feldstein, "The Global Expansion of AI Surveillance," Carnegie Endowment for International Peace, 2019; David Perez-Des Rosiers, "AI Application in Surveillance for Public Safety: Adverse Risks for Contemporary Societies," in *Towards an International Political Economy of Artificial Intelligence*, ed. Tugrul Keskin and Ryan David Kiggins (Springer, 2021), 113–43.

000 **automating processes that previously required input from humans like Petrov**: Ord, "Anthropogenic Risks"; Robert Sparrow, "Robots and Respect: Assessing the Case against Autonomous Weapon Systems," *Ethics & International Affairs* 30, no. 1 (April 2016): 93–116.

CHAPTER 6: WE MIGHT BE AFFECTING MANY BEINGS

000 **heat and humidity in Kansas**: Bill Chappell, "Days of Intense Heat Have Killed Thousands of Cattle in Kansas," NPR, June 16, 2022.

000 **heat wave in the Pacific Northwest**: Rachel H. White et al., "The Unprecedented Pacific Northwest Heatwave of June 2021," Nature Communications, February 9, 2023, 727.

000 **floods in the same region**: Chuqiao Lai et al., "Causes and Effects of the November 2021 Pacific Northwestern Floods in British Columbia," Highlights in Science, Engineering and Technology, November 10, 2022, 75–85.

000 **destroy excess inventory**: Sebo, "Animals, Pandemics, and Climate Change."

000 **human violence and neglect**: Sebo.

000 **our moral faculties are outdated**: Dale Jamieson, "The Frontiers of Ethics," in *Reason in a Dark Time* (Oxford University Press, 2014).

000 **a dominant influence on the planet**: Balter, "Archaeologists Say the 'Anthropocene' Is Here"; Lewis and Maslin, "Defining the Anthropocene"; Alice Crary and Lori Gruen, *Animal Crisis: A New Critical Theory* (John Wiley & Sons, 2022); Sebo, *Saving Animals, Saving Ourselves*; Zalasiewicz et al., "The Anthropocene."

Some scholars prefer the term "Capitalocene" to "Anthropocene" because they see human economic activity in particular, rather than human activity in general, as the dominant force on the planet. See: Jason Moore, ed., *Anthropocene or Capitalocene? Nature, History, and the Crisis of Capitalism* (Kairos, 2016).

000 **carbon-based and silicon-based beings**: For discussion of chimeras, see Hyun, "What's Wrong with Human/Nonhuman Chimera Research?" For discussion of artificial intelligence, see Gunkel, "The Other Question."

000 **potential to affect nearly everyone**: Sebo, *Saving Animals, Saving Ourselves.*

000 **a responsibility to reduce and repair this harm where possible**: Stephen Gardiner, "Accepting Collective Responsibility for the Future," *Journal of Practical Ethics* 5, no. 1 (2017).

000 **morally significant too**: For discussion of human complicity in biodiversity loss in the Anthropocene, see: Elizabeth Kolbert, *Under a White Sky: The Nature of the Future* (Crown, 2021).

000 **distinctive needs and vulnerabilities**: For more, see Elisabeth H. Ormandy, Julie Dale, and Gilly Griffin, "Genetic Engineering of Animals: Ethical Issues, Including Welfare Concerns," *Canadian Veterinary Journal* 52, no. 5 (2011): 544.

000 **as these technologies develop**: For discussion of the prospect of advanced artificial intelligence, see: Nick Bostrom, *Superintelligence: Paths, Dangers, Strategies* (Oxford University Press, 2014).

000 **like pigs or robots**: Jeff Sebo and Brendan Parent, "Human, Nonhuman, and Chimeric Research: Considering Old Issues with New Research," *Hastings Center Report* 52 (2022): S29–33.

000 **making many animals at least partly captive**: Nicolas Delon, "Pervasive Captivity and Urban Wildlife," *Ethics, Policy & Environment* 23, no. 2 (2020): 123–43.

000 **many other beings better in the future**: Sebo, "Animals, Creation Ethics, and Population Ethics."

000 **the UN Food and Agriculture Organization describes One Health**: "One Health," Food and Agriculture Organization of the United Nations, accessed November 2023.

000 **one hundred billion captive animals**: Amber Pariona, "What Is the Environmental Impact of the Fishing Industry?," World Atlas, April 25, 2017.

000 **prioritize efficiency over welfare**: Temple Grandin and Martin Whiting, eds., *Are We Pushing Animals to Their Biological Limits? Welfare and Ethical Implications* (CABI, 2018); Timothy Pachirat, *Every Twelve Seconds: Industrialized Slaughter and the Politics of Sight* (Yale University Press, 2011).

000 **animal agriculture also contributes to deforestation**: Pachirat, *Every Twelve Seconds.*

000 **14.5% of global human-caused greenhouse gas emissions**: Gerber et al., "Tackling Climate Change through Livestock."

000 **phasing up plant-based alternatives**: For more on the need to transition from animal agriculture to plant-based agriculture, see: Charlotte Blattner, "Just Transition for Agriculture? A Critical Step in Tackling Climate Change," *Journal of Agriculture, Food Systems, and Community Development* 9, no. 3 (2020): 53–58; Matthew N. Hayek, "Improving Pulse Production for a Sustainable Food Future," NYU Center for Environmental and Animal Protection, 2020; Adrian Rorheim et al., "Cultured Meat: An Ethical Alternative to Industrial Animal Farming," Sentience Politics, 2016, 1–14.

000 **care for other animals in the future**: For more on the need to build a broad, pluralistic animal advocacy movement for the future, see Jeff Sebo and Peter Singer, "Activism," in *Critical Terms for Animal Studies*, ed. Lori Gruen (University of Chicago Press, 2018), 33–45; Jeff Sebo, "Effective Animal Advocacy," in *The Routledge Handbook of Animal Ethics*, ed. Bob Fischer (Routledge, 2019), 530–42.

000 **human and nonhuman oppressions are linked**: For a more detailed discussion of some of these links, see Carol J. Adams, *The Sexual Politics of Meat: A Feminist-Vegetarian Critical Theory* (1990; Bloomsbury, 2015); Lori Gruen, "The Faces of Animal Oppression," in *Dancing with Iris: The Philosophy of Iris Marion Young*, ed. Ann Ferguson and Mechthild Nagel (Oxford University Press, 2009), 225–37; Aph Ko and Syl Ko, *Aphro-ism: Essays on Pop Culture, Feminism, and Black Veganism from Two Sisters* (Lantern Publishing & Media, 2017); Sunaura Taylor, *Beasts of Burden* (The New Press, 2017).

000 **a vast number of virtual beings for research, education, or entertainment**: For more on virtual worlds, see David Chalmers, *Reality+: Virtual Worlds and the Problems of Philosophy* (W. W. Norton, 2022).

000 **we risk amplifying human biases**: For discussion and examples of AI systems that amplify human bias, see Timnit Gebru, "Race and Gender," in *The Oxford Handbook of Ethics of AI*, ed. Markus D. Dubber, Frank Pasquale, and Sunit Das (Oxford University Press, 2019), 251–69; Thilo Hagendorff et al., "Speciesist Bias in AI: How AI Applications Perpetuate Discrimination and Unfair Outcomes against Animals," *AI and Ethics* 3, no. 3 (2023): 717–34; Adheesh Kadiresan, Yuvraj Baweja, and Obi Ogbanufe, "Bias in AI-Based Decision-Making," in *Bridging Human Intelligence and Artificial Intelligence*, ed. Mark V. Albert, Lin Lin, Michael J. Spector, and Lemoyne S. Dunn (Springer, 2022), 275–85; Justyna Stypinska, "AI Ageism: A Critical Roadmap for Studying Age Discrimination and Exclusion in Digitalized Societies," *AI & Society* 38, no. 2 (2023): 665–77.

000 **questions about AI safety and alignment will be increasingly pressing**: Dario Amodei et al., "Concrete Problems in AI Safety," arXiv, 2016; Yoshua Bengio et al., "Managing Extreme AI Risks amid Rapid Progress," arXiv, 2024; Bostrom, *Superintelligence*; Carlsmith Joseph, "Is Power-Seeking AI an Existential Risk?," arXiv, 2022; Richard Ngo, Lawrence Chan, and Sören Mindermann, "The Alignment Problem from a Deep Learning Perspective," arXiv, 2023.

000 **into paperclips**: The possibility of a "paperclip maximizer" was introduced by philosopher Nick Bostrom. Nick Bostrom, "Ethical Issues in Advanced Artificial Intelligence," in *Science Fiction and Philosophy: From Time Travel to Superintelligence*, ed. Susan Schneider (Wiley-Blackwell, 2003), 277–84.

000 ***all* the potential stakeholders of our practices**: For arguments that we should ensure that AI systems are safe for nonhuman animals as well as humans, see: Oliver Bendel, "Towards Animal-Friendly Machines," *Paladyn, Journal of Behavioral Robotics* 9, no. 1 (2018): 204–13; Joshua C. Gellers, *Rights for Robots* (Routledge, 2020); Peter Singer and Yip Fai Tse, "AI Ethics: The Case for Including Animals," *AI and Ethics* 3, no. 2 (2023): 539–51; Soenke Ziesche, "AI Ethics and Value Alignment for Nonhuman Animals," *Philosophies* 6, no. 2 (2021): 31.

CHAPTER 7: AGAINST HUMAN EXCEPTIONALISM

000 **ineligible for a human heart**: Associated Press, "A Man Who Got the 1st Pig Heart Transplant Has Died after 2 Months," NPR, March 9, 2022.

000 **porcine hearts and human bodies**: Roni Caryn Rabin, "In a First, Man Receives a Heart from a Genetically Altered Pig," *New York Times*, January 10, 2022.

000 **"We were . . . strong will to live"**: Deborah Kotz, "Study Sheds Light on Death of Pig-Heart Transplant Patient," press release, University of Maryland School of Medicine, June 28, 2022.

000 **surgery an early success**: Deborah Kotz, "University of Maryland School of Medicine Faculty Scientists and Clinicians Publish Findings of World's First Successful Transplant of Genetically Modified Pig Heart into Human Patient," press release, University of Maryland School of Medicine, June 22, 2022.

000 **carry more ethical weight**: Arthur Caplan and Brendan Parent, "The Ethics of Saving Lives through a New Type of Organ Transplant," *New York Daily News*, October 25, 2021.

000 **our responsibility to support them will increase as well**: This chapter is based on an essay I published in Aeon: Jeff Sebo, "Human Exceptionalism Is a Danger to All, Human and Nonhuman," Aeon, May 5, 2022.

000 **than nonhumans**: Kagan, *How to Count Animals.*

000 **significant bonds within our species**: Brody, "Defending Animal Research."

000 **humans have about four hundred thousand times as many neurons**: Roberto Lent et al., "How Many Neurons Do You Have? Some Dogmas of Quantitative Neuroscience under Revision," *European Journal of Neuroscience* 35, no. 1 (2012): 1–9; Takashi Shimada et al., "Analysis of the Distribution of the Brain Cells of the Fruit Fly by an Automatic Cell Counting Algorithm," *Physica A: Statistical Mechanics and Its Applications* 350, no. 1 (2005): 144–49.

000 **average fruit fly lifespan is about eighty days**: Matthew D. W. Piper and Linda Partridge, "Protocols to Study Aging in Drosophila," in *Drosophila: Methods and Protocols*, ed. Christian Dahmann (Humana Press, 2016): 291–302.

000 **more at stake in life**: The idea that death is bad because it prevents future satisfaction and happiness is sometimes called the "deprivation account" of death. See: Fred Feldman, "Some Puzzles about the Evil of Death," *Philosophical Review* 100, no. 2 (1991): 205–27; Thomas Nagel, "Death," *Noûs* 4, no. 1 (1970): 73–80.

000 **three times as many neurons as humans**: Suzana Herculano-Houzel et al., "The Elephant Brain in Numbers," *Frontiers in Neuroanatomy* 8 (2014).

000 **jellyfishes have indefinite lifespans**: Ronald S. Petralia, Mark P. Mattson, and Pamela J. Yao, "Aging and Longevity in the Simplest Animals and the Quest for Immortality," *Ageing Research Reviews* 16 (2014): 66–82.

000 **ratio of at least one hundred million to one**: Approximately 8 billion humans multiplied by 100 million is about 800 quadrillion (or 80^{17}). Eight hundred

quadrillion is a conservative estimate for the number of nonhuman animals alive at any given time, since there are conservatively 24^{10} animals in farms, 10^8 animals in labs, 10^{11} wild mammals, 10^{11} wild birds, 20^{11} reptiles and amphibians, 10^{13} fishes, and hundreds of quadrillions to tens of quintillions of arthropods. See: Brian Tomasik, "How Many Wild Animals Are There?," Reducing Suffering, 2009.

000 **would be too demanding**: Joseph Heath, *Philosophical Foundations of Climate Change Policy* (Oxford University Press, 2021).

000 **our species is still at an early stage**: William MacAskill, *What We Owe The Future* (Simon & Schuster, 2022).

000 **and the wildlife trade**: Jeff Sebo, "Methods of Inclusion of Animals," in *Saving Animals, Saving Ourselves*, 91–115.

000 **with both AI safety and AI welfare in mind**: Robert Long, Jeff Sebo, and Toni Sims, "Is AI Safety Compatible with AI Welfare?" (unpublished manuscript).

CONCLUSION: THINK COSMICALLY, ACT GLOBALLY

000 **thought experiment popularized by John Rawls**: John Rawls, *A Theory of Justice*, rev. ed. (1971; Harvard University Press, 1999); John Rawls, *Political Liberalism*, 2nd ed. (1993; Columbia University Press, 2005).

000 **in such societies**: Rawls describes cooperation as one of the "circumstances of justice" in his book *A Theory of Justice.*

000 **to the global context**: John Rawls, *The Law of Peoples* (Harvard University Press, 1999).

000 **to the multispecies context, too**: For an overview, see Alasdair Cochrane, *An Introduction to Animals and Political Theory* (Palgrave Macmillan, 2010).

000 **reflective equilibrium**: Reflective equilibrium is a method of forming well-considered judgments from our intuitions. It was developed as a method of moral and political philosophy but has been extended to other areas. See: Norman Daniels, *Justice and Justification: Reflective Equilibrium in Theory and Practice* (Cambridge University Press, 1996); Nelson Goodman, *Fact, Fiction, and Forecast*, 2nd ed. (Bobbs-Merrill, 1965); Rawls, *Theory of Justice.*

000 **subject language**: Adams, *Sexual Politics of Meat.*

000 **global change requires local action**: Daniel Tarantola, "Thinking Locally, Acting Globally?," *American Journal of Public Health* 103, no. 11 (2013).

INDEX

Norton Shorts

BRILLIANCE WITH BREVITY

W. W. Norton & Company has been independent since 1923, when William Warder Norton and Mary (Polly) D. Herter Norton first published lectures delivered at the People's Institute, the adult education division of New York City's Cooper Union. In the 1950s, Polly Norton transferred control of the company to its employees.

One hundred years after its founding, W. W. Norton & Company inaugurates a new century of visionary independent publishing with Norton Shorts. Written by leading-edge scholars, these eye-opening books deliver bold thinking and fresh perspectives in under two hundred pages.

Available Winter 2025

Imagination: A Manifesto by Ruha Benjamin

What's Real About Race?: Untangling Science, Genetics, and Society by Rina Bliss

Offshore: Stealth Wealth and the New Colonialism by Brooke Harrington

Fewer Rules, Better People: The Case for Discretion by Barry Lam

Explorers: A New History by Matthew Lockwood

Wild Girls: How the Outdoors Shaped the Women Who Challenged a Nation by Tiya Miles

The Moral Circle: Who Matters, What Matters, and Why by Jeff Sebo

Against Technoableism: Rethinking Who Needs Improvement by Ashley Shew

Literary Theory for Robots: How Computers Learned to Write by Dennis Yi Tenen

Forthcoming

Mehrsa Baradaran on the racial wealth gap

Merlin Chowkwanyun on the social determinants of health

Daniel Aldana Cohen on eco-apartheid

Jim Downs on cultural healing

Reginald K. Ellis on Black education versus Black freedom

Nicole Eustace on settler colonialism

Agustín Fuentes on human nature

Justene Hill Edwards on the history of inequality in America

Destin Jenkins on a short history of debt

Quill Kukla on a new vision of consent

Kelly Lytle Hernández on the immigration regime in America

Natalia Molina on the myth of assimilation

Rhacel Salazar Parreñas on human trafficking

Tony Perry on water in African American culture and history

Beth Piatote on living with history

Ashanté Reese on the transformative possibilities of food

Tracy K. Smith on poetry in an age of technology

Daniel Steinmetz-Jenkins on religion and populism

Onaje X. O. Woodbine on transcendence in sports